It's another Quality Book from CGP

This book is for anyone doing GCSE Maths at Foundation Level.

It contains lots of tricky questions designed
to make you sweat — because that's the only
way you'll get any better.

It's also got some daft bits in to try and make
the whole experience at least vaguely
entertaining for you.

What CGP is all about

Our sole aim here at CGP is to produce the highest quality
books — carefully written, immaculately presented and
dangerously close to being funny.

Then we work our socks off to get them out to you
— at the cheapest possible prices.

Contents

Section Four

Statistics and Graphs

Section Five

Angles & Other Bits

Section Six

Algebra

Section Seven

Exam Questions

Published by Coordination Group Publications Ltd.
Illustrated by Ruso Bradley, Lex Ward and Ashley Tyson

Coordinated by June Hall and Mark Haslam

Contributors:
Philip Wood
Margaret Carr
Barbara Coleman
John Lyons
Gordon Rutter
Claire Thompson

ISBN 1 84146 019 2

Groovy website: www.cgpbooks.co.uk

Printed by Elanders Hindson, Newcastle upon Tyne.
Clipart sources: CorelDRAW and VECTOR.

1.1   Questions on Writing Numbers

Q1   Write these words as numbers.

a) Twenty six   ....

b) Seventy three   ....

c) Eight hundred and sixty   ....

d) Five thousand seven hundred and ninety two   ....

e) Twenty seven thousand and fifty two   ....

f) Four hundred and sixty three thousand and four   ....

g) Mr U.N.Lucky won three hundred and fourteen pounds on the Lottery. Write
   this in the space on his cheque.

His friend Mr L.Ucky won four million six hundred and seventy three thousand
five hundred and twelve pounds on the lottery. Write this amount in the space
on his cheque.

Q2   Write these numbers as words.

a) 27   ....

b) 507   ....

c) 3,824   ....

d) 63,492   ....

e) 245,094   ....

f) 2,172,603   ....

*Look at the big numbers in groups of 3 digits — start from the right and work
left, putting a comma in front of every 3 digits you write.*

1.2   *Questions on Ordering Numbers*

Q1   What is the value of the 5 in each of these numbers? Write your answer in words.

     a) 153   ....   **b)** 25   ....

     c) 6523   ....   **d)** 75,392   ....

Q2   Put these numbers in ascending, smallest to biggest, order.

     a) 23   117   5   374   13   89   67   54   716   18

     ....

     b) 1272   231   817   376   233   46   2319   494   73   1101

     ....

     c) Five friends took part in a darts tournament.
        The table shows the total number of points scored by each friend.
        Re-write the table showing who got most to who got least points.

Glen	2317
Sharon	854
Jason	1179
Melissa	1873
Paul	724

Q3   Put these numbers in descending, biggest to smallest, order.

     a) 323   16   98   514   5   18   429   25   87   11

     ....

     b) 45   286   71   6451   95   6295   762   245   39   7164

     ....

FIRST put them in groups of how many **DIGITS** there are — **THEN**, and **ONLY THEN** should you start putting each group in order.

1.3 Questions on Adding without the Calculator

 ALWAYS put the numbers in **COLUMNS** when you're adding... and check the **UNITS, TENS & HUNDREDS** line up.

Q1 Do these questions as quickly as you can, writing the answers in the spaces provided:

a) 5 + 9= ....

b) 26 + 15= ....

c) 34 + 72= ....

d) 238 + 56= ....

e) 528 + 173= ....

f) 215 + 2514= ....

Q2 Now try these:

a) 63
　+32

b) 75
　+48

c) 528
　+196

Q3 What is the missing number?

a) 37 + 　 = 89

b) 63 + 　 = 92

c) 236 + 　 = 305

Q4 Add the rows and add the columns:

a)

2	6	7	
8	6	4	
4	2	9	

b)

8	3	7	
2	6	9	
7	3	4	

c)

2	7	1	
6	5	8	
3	6	2	

Q5 These boxes in a music warehouse contain CD's.

62	218	894	42
361	1283	59	732
54	745	29	319

Select the three boxes that contain the most CD's.
How many CD's will you have?
Put your answer in the shaded box.

			Total

Select the three boxes that contain the least number of CD's.
How many do you have now?
Put your answer in the shaded box.

			Total

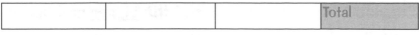

SECTION ONE — NUMBERS MOSTLY

1.4   *Questions on Subtracting*

Q1   Subtract the following <u>without</u> a calculator:

a) 36 – 13 = ...   b) 45 – 23 = ...

c) 89 – 24 = ...   d) 25 – 8 = ...

e) 80 – 42 = ...   f) 72 – 19 = ...

g)   687
    –235

h)   754
    –538

i)   843
   –516

j)   634
   – 98

k)   908
   –325

l)   650
   –317

m)   830
   – 293

n)   700
   – 248

o)   6359
   –2263

p)   4173
   –1324

q)   5271
   –2481

r)   4000
   – 1692

Q2   Fill in the missing digits:

a)   6 5
  –3☐
  ☐4

b)   7 3☐
  –2☐4
  ☐2 5

c)   8 7☐
  –☐3 2
  2 ☐9

d)   ☐56
  –2 78
  1 ☐☐

Q3   At the beginning of the day a supermarket had 462 tins of beans.
By the end of the day 345 had been sold. How many were left? ....

Q4   Scafell Pike is 979m high. Ben Nevis is 1344m high. What is the difference in
height between the two mountains?

....

    Yeah, I know it's obvious, but <u>remember</u> to put the <u>bigger</u> number at the
<u>top</u> before you start subtracting.

1.5   *Questions on Multiplying*

Don't forget to put a zero under the units when you multiply by that extra number in the tens column.

Q1   Multiply the following <u>without</u> a calculator:

a) 23 × 2 = ...   b) 40 × 3 = ...   c) 53 × 4 = ...

d) 13 × 5 = ...   e) 25 × 4 = ...   f) 42 × 3 = ...

g)   18
     × 2

h)   54
     × 3

i)   75
     × 5

j)   93
     × 4

k)   308
     × 4

l)   825
     × 3

m)   346
     × 5

n)   286
     × 6

o)   126
     × 14

p)   413
     × 26

q)   309
     × 61

r)   847
     × 53

Q2   What is the total cost of 6 pens at 54p each?

....

Q3   How many hours are there in a year (365 days)?

....

1.6   *Questions on Dividing*

Q1   Do these divisions without a calculator:

a) 46 ÷ 2   You may wish to set the sum out like this   2)‾23‾/46

b) 86 ÷ 2   c) 96 ÷ 3   d) 76 ÷ 4

e) 85 ÷ 5   f) 96 ÷ 6   g) 91 ÷ 7

Q2   Try these:

a) 834 ÷ 3   b) 645 ÷ 5   c) 702 ÷ 6

d) 900 ÷ 4   e) 1000 ÷ 8   g) 747 ÷ 9

Q3   Seven people share a lottery win of £868.00. How much did each person get?

Q4   A chocolate cake containing 944 calories is split into 8 slices. How many calories are in each slice?

OK they're easier with the calulator, but even without it they're a bit of a doddle — just make sure you set the sum out neatly.

1.7   Questions on Writing Decimals

See — decimals aren't so bad after all, are they...

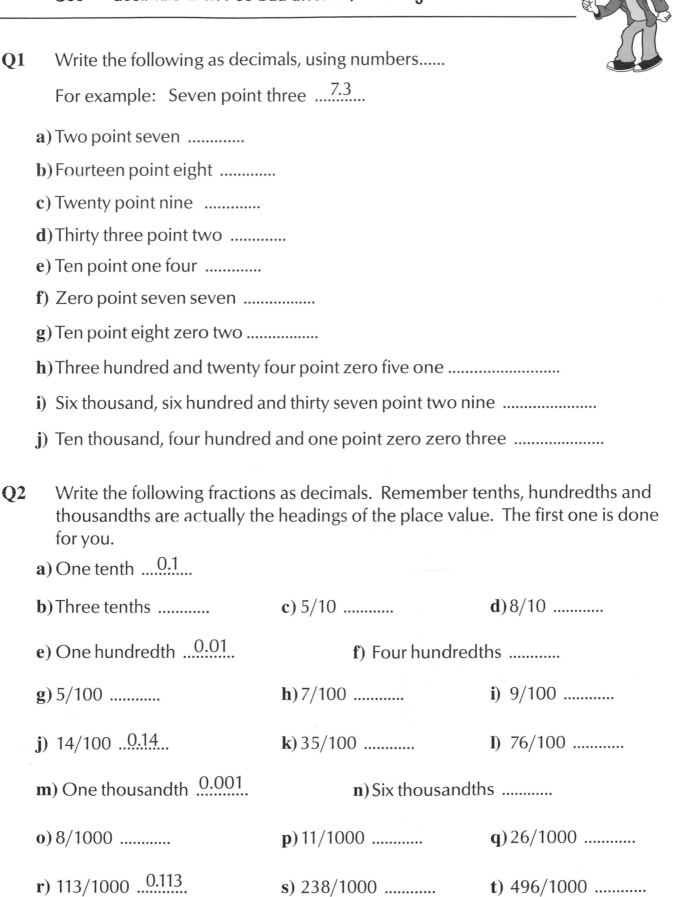

Q1   Write the following as decimals, using numbers......

For example: Seven point three .....7.3.....

a) Two point seven ....

b) Fourteen point eight ....

c) Twenty point nine ....

d) Thirty three point two ....

e) Ten point one four ....

f) Zero point seven seven ....

g) Ten point eight zero two ....

h) Three hundred and twenty four point zero five one ....

i) Six thousand, six hundred and thirty seven point two nine ....

j) Ten thousand, four hundred and one point zero zero three ....

Q2   Write the following fractions as decimals. Remember tenths, hundredths and thousandths are actually the headings of the place value. The first one is done for you.

a) One tenth .....0.1.....

b) Three tenths ....   c) 5/10 ....   d) 8/10 ....

e) One hundredth ....0.01....   f) Four hundredths ....

g) 5/100 ....   h) 7/100 ....   i) 9/100 ....

j) 14/100 ....0.14....   k) 35/100 ....   l) 76/100 ....

m) One thousandth ....0.001....   n) Six thousandths ....

o) 8/1000 ....   p) 11/1000 ....   q) 26/1000 ....

r) 113/1000 ....0.113....   s) 238/1000 ....   t) 496/1000 ....

1.8 Questions on Ordering Decimals

Look at the numbers before the point first, and do what you did before... then work right from the decimal point, comparing each number in turn.

Q1 John has £2.12, Philip has £2.21 and Gordon has £2.02. Who has the most money?

....

Q2 Lisa's time for a sprint race was 12.32 seconds. Kate's time was 12.34 seconds. Who ran fastest ?

....

Q3 Circle the larger number in each pair.

0.64   0.46   0.05   0.5   15.83   15.89   1.7216   1.7126

Q4 > means "is greater than"   < means "is less than". Write "true" or "false" in each case:

0.63 > 0.36 ....;   0.06 > 0.055 ....;   8.2 < 8.022 ....

Q5

Place these numbers at A, B and C on the number line: 2.55   2.5   2.05

Q6

Place these numbers at A, B, C and D on the number line:
0.123   0.321   0.312   0.213

Q7 Put these numbers in order, from the smallest to the largest.

a) 3.42   4.23   2.43   3.24   2.34   4.32

....

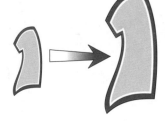

b) 6.7   6.704   6.64   6.642   6.741

....

c) 1002.8   102.8   1008.2   1020.8   108.2

....

1.9   *Questions on Adding Decimals*

Q1   Work out the answers without using a calculator.

a)   2.4
    + 3.2

b)   3.5
    + 4.6

c)   6.2
    + 5.9

d)   7.34
    + 6.07

e)   9.08
    + 4.93

f)   15.73
    + 25.08

g)   26.05
    + 72.95

Q2   Write these out in columns and work out the answer without using a calculator.

a) 3.6 + 7.3   **b)** 21.4 + 13.8   **c)** 0.9 + 5.6   **d)** 9.98 + 6.03   **e)** 2.9 + 7

f) 4.36 + 7.1   **g)** 9.8 + 1.05   **h)** 6 + 6.75   **i)** 0.28 + 18.5   **j)** 47.23 + 6.7

Q3   Work out the missing lengths.

a)

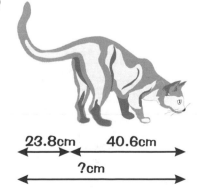

23.8cm   40.6cm

?cm

b)

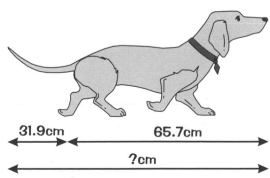

31.9cm   65.7cm

?cm

Some questions don't have the decimal point, so you'll have to put it in yourself — you'll have to add some zeros, too... careful now.

1.10   Questions on Subtracting Decimals

Q1   Work these out <u>without</u> a calculator:

a) 9.6 – 4.3 = ...   **b)** 10.8 – 3.5 = ...   **c)** 8.4 – 6.4 = ...

d)   9.8   **e)**   7.3   **f)**   6.2   **g)**   8.6
        – 3.1   – 2.3   – 1.5   – 3.9

h)   7.0   **i)**   13.6   **j)**   14.65   **k)**   8.34
        – 1.6   – 12.7   – 4.7   – 4.65

Q2   Put the following in columns first then work them out:

a) 8.5 – 1.6   **b)** 18.3 – 5.9   **c)** 24.1 – 16.3

d) 9 – 3.6   **e)** 40 – 2.3   **f)** 51 – 18.32

Q3   Kate bought a jar of coffee for £3.24 and paid for it with a £5 note. How much change should she get?

....

Q4   Work out the height of the table which the television is standing on.

1.18m   0.48m   ?

....

When you're dealing with MONEY, you've only got TWO decimal places.

1.11 Questions on Multiplying Decimals

Ignore the decimal point to start with — just multiply the numbers.
Then put the point back in and **CHECK** your answer sounds sensible.

Q1 **a)** 3.2 × 4 = ...   **b)** 8.3 × 5 = ...   **c)** 6.4 × 3 = ...

d) 21 × 0.3 = ...   **e)** 35 × 0.4 = ...   **f)** 263 × 0.2 = ...

g) 2.4 × 3.1 = ...   **h)** 5.3 × 2.4 = ...   **i)** 1.7 × 6.8 = ...

Q2   Jason can run 5.3 metres in 2 seconds. How far will he run if he keeps up this pace for:

a) 20 seconds   **b)** 60 seconds   **c)** 5 minutes

....

Carl can run 7.4 metres in 2 seconds. How far will he run if he keeps up this pace for:

d) 20 seconds   **e)** 60 seconds   **f)** 5 minutes

....

g) If they both keep up their pace for half an hour how far will each of them run?

Jason will run .... metres.

Carl will run .... metres.

Q3   At a petrol station each pump shows a ready reckoner table. Complete the table when the cost of unleaded petrol is 64.9p per litre.

Litres	Cost in pence
1	64.9
5	
10	
20	
50	

1.12   *Questions on Dividing Decimals*

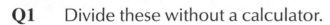 This is the same thing, really — you divide the numbers first, then you put in the point where it should be.

no calculators!!

Q1   Divide these without a calculator.

a) 8.4 ÷ 2   You may wish to set the sum out like this

$$\begin{array}{r} 4.2 \\ 2\,\overline{)\,8.4} \end{array}$$

b) 7.5 ÷ 3   **c)** 8.5 ÷ 5   **d)** 26.6 ÷ 7

e) 4.75 ÷ 5   **f)** 8.28 ÷ 9   **g)** 0.944 ÷ 8

Q2   Try these:

a) 6.2 ÷ 5   **b)** 2.3 ÷ 4   **c)** 0.9 ÷ 5

d) 2.06 ÷ 8   **e)** 0.405 ÷ 6   **f)** 0.3 ÷ 8

Q3   Eight people share £9.36. How much does each get?

....

Q4   A plank of wood 8.34m long is cut into 6 equal pieces. How long is each?

....

1.13   Questions on Place Value

Q1   Write down the value of the number 4 in each of these...

For example 408 .....*hundreds*.....

a) 347 ....   b) 41 ....   c) 5478 ....

d) 6754 ....   e) 4897 ....   f) 6045 ....

g) 64098 ....   h) 745320 ....   i) 405759 ....

j) 2402876 ....   k) 4987321 ....   l) 6503428 ....

Q2   Write the value of the number 2 in the following...

For example 98.2 .....*tenths*.....

a) 832.7 ....   b) 30.25 ....   c) 523.68 ....

d) 134.20 ....   e) 9824.57 ....   f) 2035.407 ....

g) 3643.023 ....   h) 752.368 ....   i) 0.352 ....

j) 563.002 ....   k) 35.026 ....   l) 12.36 ....

Q3   Using the digits 0, 1, 2, 3 and 4 once only, write down a five-digit number that has...

a) 3 in the units ....   b) 4 in the thousands ....

c) 2 in the tens ....   d) 1 in the hundreds ....

e) 0 in the tenths ....

Q4   Rearrange the digits 0, 1, 2, 3, 4 and 5 to make the biggest and smallest whole numbers possible.

Biggest ....   Smallest ....

Putting in the commas will make it easier to see what's what.

1.14 Questions on Multiplying by 10, 100, etc.

Fill in the missing numbers. Do not use a calculator for this page.

Q1 $6 \times \boxed{} = 60$ $0.07 \times \boxed{} = 0.7$

  $6 \times \boxed{} = 600$ $0.07 \times \boxed{} = 7$

  $6 \times \boxed{} = 6000$ $0.07 \times \boxed{} = 70$

Q2   Which number is ten times as large as 25?   ....

Q3   Which number is one hundred times as large as 93?   ....

Q4 $8 \times 10 =$ .... $34 \times 100 =$ .... $52 \times 100 =$ ....

  $9 \times 1000 =$ .... $436 \times 1000 =$ .... $0.2 \times 10 =$ ....

  $6.9 \times 10 =$ .... $4.73 \times 100 =$ ....

  $0.65 \times 1000 =$ .... $3.7 \times 1000 =$ ....

Q5   For a school concert chairs are put out in rows of 10. How many will be needed for 16 rows? ....

Q6      How much do 10 chickens cost? ....

£2.99 EACH

Q7   A shop bought 1000 bars of chocolate for £0.43 each.
How much did they cost altogether?

  ....

Q8      A school buys calculators for £2.45 each.
How much will 100 cost? ....

Q9 $20 \times 30 =$ .... $40 \times 700 =$ .... $250 \times 20 =$ ....

  $6000 \times 210 =$ .... $18000 \times 500 =$ ....

Multiplying by 10, 100 or 1000 moves each digit 1, 2 or 3 places to the left —
you just fill the rest of the space with zeros.

1.15 Questions on Dividing by 10, 100, etc.

Keep track of which way that decimal point's moving.

Q1 Work these out <u>without</u> a calculator:

a) 30 ÷ 10 = ...

b) 43 ÷ 10 = ...

c) 5.8 ÷ 10 = ...

d) 63.2 ÷ 10 = ...

e) 0.5 ÷ 10 = ...

f) 400 ÷ 100 = ...

g) 423 ÷ 100 = ...

h) 228.6 ÷ 100 = ...

i) 61.5 ÷ 100 = ...

j) 2.96 ÷ 100 = ...

k) 6000 ÷ 1000 = ...

l) 6334 ÷ 1000 = ...

m) 753.6 ÷ 1000 = ...

n) 8.15 ÷ 1000 = ...

o) 80 ÷ 20 = ...

p) 860 ÷ 20 = ...

q) 2400 ÷ 300 = ...

r) 480 ÷ 40 = ...

s) 860 ÷ 200 = ...

t) 63.9 ÷ 30 = ...

Q2 Ten people share a Lottery win of £62.
How much should each person receive?

....

Q3 Blackpool Tower is 158m tall. If a model
of it is built to a scale of 1 : 100 , how tall
would the model be?

....

Q4 If 1000 identical ball-bearings weigh 2100g,
what is the weight of one of the ball-bearings?

....

Q5 Mark went on holiday to France. He exchanged £100 for 891 francs to spend
while he was there. How many francs did he get for each £1 ?

....

1.16   Questions on Odds and Evens

Even numbers divide by 2.

Odd numbers do not divide by 2. Easy, innit.

Q1   Look at the list of the counting numbers from 1 to 20. Circle the numbers that divide by 2. Put a box around the numbers that do not divide by 2.

1   2   3   4   5   6   7   8   9   10

11   12   13   14   15   16   17   18   19   20

Make a list of the next ten even numbers.

....

Make a list of the next ten odd numbers.

....

Q2   Number bugs. Look at the number bug. Can you see how it grows? If the body part is an even number divide by 2. If the body part is an odd number then add 1.

Now fill in the numbers on the body parts.

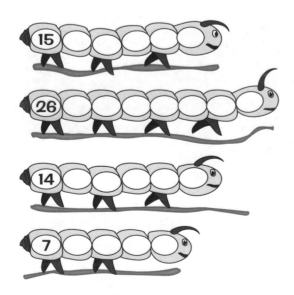

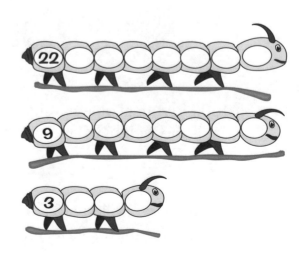

1.17 Questions on Multiples

Q1   What are the first five multiples of:

a) 4 ....

b) 7 ....

c) 12 ....

d) 18 ....

Q2   Find a number which is a multiple of:

a) 2 and 6 ....

b) 7 and 5 ....

c) 2 and 3 and 7 ....

d) 4 and 5 and 9 ....

Q3 a) Find a number which is a multiple of 3 and 8 ....

b) Find another number which is a multiple of 3 and 8 ....

c) Find another number which is a multiple of 3 and 8 ....

Q4   Which of these number 14, 20, 22, 35, 50, 55, 70, 77, 99 are multiples of:

a) 2 ....

b) 5 ....

c) 7 ....

d) 11 ....

The multiples of a number are its times table — if you need multiples of more than one number, do them separately then pick the ones in both lists.

1.18   *Questions on Factors*

Factors multiply together to make other numbers

Eg $1 \times 6 = 6$ and $2 \times 3 = 6$, so 6 has factors 1, 2, 3 and 6.

Q1 These are the hands of Aliens. Each Alien has a times sum on each finger. The answer to each finger is the same and is the number by which that Species is known.

Species 8   Species 12   Species 7

When Sculder & Mulley found the following Aliens they had all their fingers but were missing their factor sums. Write in each sum and the Alien Species number.

Species ___   Species 30   Species ___

Q2 The numbers on each finger are known as factors and are usually, on Earth, written as a list. List the factors of the following numbers. Each factor is written once, no repeats.

a) 18 ....

b) 22 ....

c) 35 ....

d) 7 ....

e) 16 ....

f) 49 ....

g) 48 ....

h) 31 ....

i) 50 ....

j) 62 ....

k) 81 ....

l) 100 ....

1.18 Questions on Factors

Q3 **a)** I am a factor of 24.
I am an odd number.
I am bigger than 1.
What number am I?

....

b) I am a factor of 30.
I am an even number.
I am less than 5.
What number am I?

....

Q4 Circle all the factors of 360 in this list of numbers.

1   2   3   4   5   6   7   8   9   10

Q5 A perfect number is one where the factors add up to the number itself.
For example, the factors of 28 are 1, 2, 4, 7 and 14 (not including 28 itself).
These add up to 1+2+4+7+14 = 28, and so 28 is a perfect number.

Complete this table, and circle the perfect number in the left hand column.

Number	Factors	Sum of Factors
2		
4	1, 2	3
6		
8		
10		

The sum of the factors is all the factors added together.

Q6 **a)** What is the biggest number that is a factor of both 42 and 18? ....

b) What is the smallest number that has both 4 and 18 as factors? ....

Q7 Fill in the missing numbers in the factor trees.
The first one has been done for you.

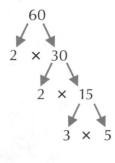

60 = 2×2×3×5

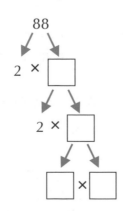

88 = 2×2×..........×..........

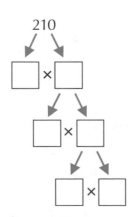

32 = ... × .... × .... × ....

SECTION ONE — NUMBERS MOSTLY

1.19 Questions on Squares and Cubes

Q1 Work out <u>without</u> a calculator:

a) $5^2 = $ .... **b)** $7^2 = $ .... **c)** $2^3 = $ .... **d)** $4^3 = $ ....

e) $6^2 = $ .... **f)** $5^3 = $ .... **g)** $9^2 = $ .... **h)** $10^3 = $ ....

Q2 Do these <u>with</u> a calculator:

a) $1.4^2 = $ .... **b)** $3.5^2 = $ .... **c)** $5.95^2 = $ .... **d)** $7.63^3 = $ ....

e) $3^2 + 6^2 = $ .... **f)** $7^3 - 5^3 = $ .... **g)** $6^3 - 14^2 = $ ....

Q3 Put a ring round the square numbers in the following list:

17   6   9   15   4   16   21   11   20   1   50

Q4 Put a ring round the cube numbers in the following list:

8   25   27   1   100   125   42   10   16   30   18

Q5 Work out:

a) 10 squared = ... **b)** 4 cubed = ...

Q6 What is the next square number after 64? ....

Q7 What is the next cube number after 216? ....

Q8 Work out with or without a calculator:

a) $2^3 \times 5^2 = $ .... **b)** $3^2 \times 1^3 = $ .... **c)** $3^3 \times 2^2 = $ ....

Q9 Write down two square numbers which are also cube numbers.

Q10 A block of flats with 20 floors has 20 windows on each floor. How many windows does the building have altogether?
If it takes 20 minutes to clean each window, how many minutes does it take to clean the block?

....

**These aren't as bad as they seem — "three squared" = $3^2 = 3 \times 3 = 9$.
Same with cubes — "ten cubed" = $10^3 = 10 \times 10 \times 10 = 1000$.**

SECTION ONE — NUMBERS MOSTLY

1.20   *Questions on Ratio*

Ratios compare quantities of the same kind — so if the units aren't mentioned, they've got to be the same in each bit of the ratio.

Q1   What is the ratio in each of these pictures?

a)

Circles to Triangles

.... to ....

Triangles to Circles

.... : ....

b)

Small stars to big stars

.... to ....

Big stars to small stars

.... : ....

Q2   Write these ratios in their simplest form. The first one is done for you.

a) 4 to 6   **b)** 15 to 21   **c)** 14 to 42   **d)** 72 to 45

   2 : 3   ...... : .... : .... : ......

e) 24 cm to 36 cm   **f)** 350 g to 2 kg   **g)** 42 p to £ 1.36

...... : .... : .... : ......

Q3   In 'The Pink Palace' The walls are painted a delicate shade of pink using 2 tins of red paint to every 3 tins of white paint.

Write this as a ratio.
Red tins .... to White tins ....

R........... : W...........

How much white paint is needed to mix with 12 red tins? .... White tins.

1.20   *Questions on Ratio*

Q4   To make grey paint, black and white paint are mixed in the ratio 5:3. How much black paint would be needed with:

a) 6 litres of white ....

b) 12 litres of white ....

c) 21 litres of white ....?

Q5   To make orange squash you mix water and concentrated orange juice in the ratio 9:2. How much water is needed with:

a) 10 ml of concentrated juice ....

b) 30 ml of concentrated juice ....

c) 42 ml of concentrated juice ....?

Q6   The ratio of men to women at a football match as 11:4. How many men were there if there were:

a) 2000 women ....

b) 8460 women ....?

How many women were there if there were:

c) 22000 men ....

d) 6820 men ....?

I think I can spot another Golden Rule lurking here...
DIVIDE FOR ONE, THEN TIMES FOR ALL.

1.20   *Questions on Ratio*

Q7   Divide the following quantities by the given ratio.

For example:

£400 in the ratio 1 : 4   1 + 4 = 5   400 ÷ 5 = 80

             1 × £80 = £80 and 4 × 80 = £320   £80 : £320

a) 100 g in the ratio 1 : 4   .... + ... = ... ÷ .... = ...

                              .... × .... = ... and .... × .... = ...

                              = ... : ....

b) 500 m in the ratio 2 : 3 = ... : ....

c) £12000 in the ratio 1 : 2 = ... : ....

d) 6.3 kg in the ratio 3 : 4 = ... : ....

e) £8.10 in the ratio 4 : 5 = ... : ....

Q8   Now try these...

a) Adam and Mags win £24 000. They split the money in the ratio 1 : 5. How much does Adam get?

....

b) Sunil and Paul compete in a pizza eating contest. Between them they consume 28 pizzas in the ratio 3 : 4. Who wins and how many did they eat?

.... eats .... pizzas

c) The total distance covered in a triathlon (swimming, cycling and running) is 15km. It is split in the ratio 2 : 3 : 5. How far is each section?

Swimming = ...   Cycling = ...   Running = ...

A great way to check your answer works is to add up the individual quantities — they should add up to the original amount.

1.21   Questions on Money

Brians Fish and Chips — MENU		
Fish and Chips — £2.10	Fish — £1.40	Sausage — 45p
Chips — 70p	Fish Cake — 35p	Curry Sauce — 50p
	Burger — £1.10	

Work out how much each order costs.

USE A CALCULATOR FOR ALL THESE QUESTIONS.

Q1   Fish and chips twice ....

Q2   Fish cake and chips ....

Q3   Fish and chips and a curry sauce ....

Q4   Burger and chips ....

Q5   Sausage and chips ....

Q6   2 Fish, 1 burger and 1 curry sauce ....

Q7   3 Sausages and a curry sauce ....

Q8   How much change do you get from £5 for:

   a) Fish and chips twice ....
      c) Burger and chips ....

   b) Sausage and chips ....

Q9   Bananas cost £1.24 per kilo. Work out the cost of

   3 kilos ....
   5 kilos ....
   2.4 kilos ....

Q10   Sausages cost £1.06 per pound. Work out the cost of:

   4 pounds ....
   5 pounds ....
   6.7 pounds ....

Q11

Shampoo £1.55	Toothpaste £1.27	Soap 63p
Deodorant £1.09	Comb 49p	Cotton Wool 35p

Work out the bills for these purchases:

Shampoo and cotton wool .... ;   Toothpaste and soap .... ;
Deodorant, comb, cotton wool and toothpaste ....

There's nothing too tricky here — it's all adding and multiplying, really.

1.22   *Questions on Best Buys*

Start by finding the AMOUNT PER PENNY — the more of the stuff you get per penny, the better value it is.

Q1   Which of these boxes of eggs is better value for money?

Grade AA 6 EGGS   Grade AA 12 EGGS

60p   £1.10   ....

Q2

The small bar of chocolate weighs 50g and costs 32p.
The large bar weighs 200g and costs 80p.

a) How many grams do you get for 1p from the small bar?

                                             ....

b) How many grams do you get for 1p from the large bar?

                                             ....

c) Which bar gives you more for your money?

                                           ....

Q3   The large tin of tuna weighs 400g and costs £1.05.
The small tin weighs 220g and costs 57p.

a) How many grams do you get for 1p in the large tin?

                                             ....

b) How many grams do you get for 1p in the small tin?

                                             ....

c) Which tin gives better value for money?

                                           ....

1.23 Questions on BODMAS

Remember this old chap — he tells you which order to work things out.
The most important bit is that brackets have priority over everything else.

Brackets   Over   Division   Multiplication   Addition   Subtraction

Q1 **a)** 4 + 3 × 2 = ...   **b)** 4 × 3 + 2 = ...   **c)** 4 – 3 × 2 = ...

d) 9 ÷ 3 + 5 = ...   **e)** 12 ÷ 4 + 5 = ...   **f)** 7 – 10 ÷ 2 = ...

Q2 **a)** 7 × (3 + 5) = ...   **b)** 6 – (8 ÷ 2) = ...   **c)** 15 ÷ (9 – 4) = ...

d) (6 + 7) × 3 = ...   **e)** (18 – 6) ÷ 3 = ...   **f)** (21 – 7) × 2 = ...

Q3 **a)** 14 – (2 + 5) = ...   **b)** (14 – 2) + 5 = ...   **c)** (14 ÷ 2) + 5 = ...

d) 20 – (10 ÷ 2) = ...   **e)** (20 ÷ 10) – 2 = ...   **f)** 20 + (10 – 2) = ...

Q4   Three cards are picked from a pack. These are: 4,   3,   6.

Put ÷,   ×,   +,   –   or ( )   in each sum to make it true.

a) 4   6   3 =   6   **b)** 4   6   3 = 30

c) 4   6   3 = 27   **d)** 4   6   3 = 13

Q5 **a)** 4 + 6 × 5 – 3 = ...   **b)** 4 × 6 + 5 × 3 = ...

c) 12 ÷ 3 + 4 × 2 = ...   **d)** 7 × 4 ÷ 2 – 3 = ...

1.24 Questions on the Calculator

Q1 In my school there are 32 classes and each class has 27 pupils. How many pupils are there in my school?

....

Q2 There are 28 sweets in a bag, together they weigh 126g. How heavy is each sweet?

....

Q3 The population of Carrsville is 6186, the population of Colesville is 9144.

a) What is the total population of the 2 towns? ....

b) What is the difference between the populations of the 2 towns? ....

Q4 If 17 pencils cost £1.53, work out the cost of:

a) 1 pencil ....

b) 5 pencils ....

c) 20 pencils ....

Q5 In Woodstown there are 12162 females and 7164 males. In Lyonstown there are 9672 females and 11471 males.

a) What is the total population of the 2 towns? ....

b) Which town is the biggest and by how many? ....

c) In total are there more males or females? By how many? ....

Q6 At the end of last week my car had travelled 26142 miles. For the last 7 days I travelled exactly 68 miles each day.

a) How far did I travel in those 7 days? ....

b) How many miles in total has my car travelled? ....

Q7 Petrol for my car costs 66.6p per litre. Find the cost of:

a) 5 litres ....   b) 20 litres ....   c) 30 litres ....

I can travel 9 miles on one litre of petrol. How far can I travel with:

d) 5 litres ....   e) 20 litres ....   f) 30 litres ....?

How much does it cost to travel:

g) 45 miles ....   h) 1 mile ....?

These aren't too bad — especially when you're allowed your calculator.
The only slightly tricky bit is pulling out the bits you need to do the sum.

2.1   Questions on Perimeters

What you've gotta do with these is add up all the sides to get the perimeter.
If there's no drawing, do it yourself — then you won't forget any of the sides.

Q1   Work out the perimeters of the following shapes :

a)

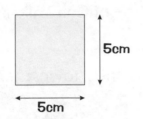

Square Perimeter = ... cm

b)

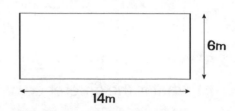

Rectangle Perimeter =
=(2 × ......) + (2 × ......) = ...m

c)

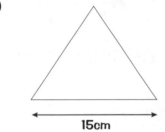

Equilateral Triangle
Perimeter = 3 × .... = ...cm

d)

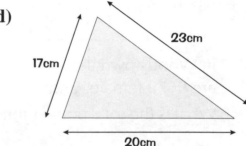

Triangle
Perimeter = ...+.......+.......=.......cm

e)

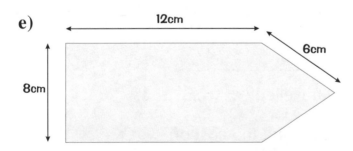

Five Sided Shape
Perimeter = ...+.......+......+.......+.......
=.......cm

f)

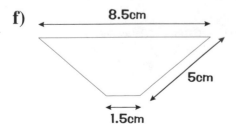

Four Sided Shape
Perimeter = ...+.......+.......+.......=.......cm

Q2 **a)** A square garden has sides of length 10m. How much fencing is needed to go around it? ....m.

b) A photo measures 17.5cm by 12.5 cm. What is the total length of the frame around it? ....cm.

2.1   *Questions on Perimeters*

Q3   Find the perimeter of these shapes (you may need to work out some of the lengths):

a)

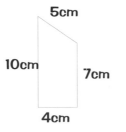

Perimeter ....

b)

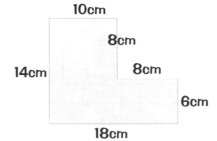

Perimeter ....

c)

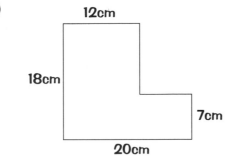

Perimeter ....

d)

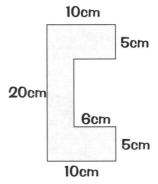

Perimeter ....

e)

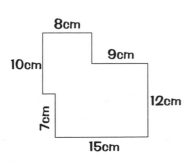

Perimeter ....

f)

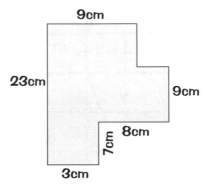

Perimeter ....

Always use the BIG BLOB METHOD — especially with more complicated shapes. It never fails — I promise...

2.2 Questions on Areas

Q1 Here are the outlines of the footprints of some animals. Each square represents 1cm². Estimate the area of each footprint:

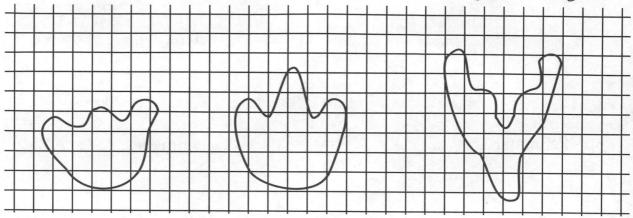

Area .... cm²   Area .... cm²   Area .... cm²

Q2 Here are the maps of 2 islands. Each square represents 1km². Estimate the area of each island:

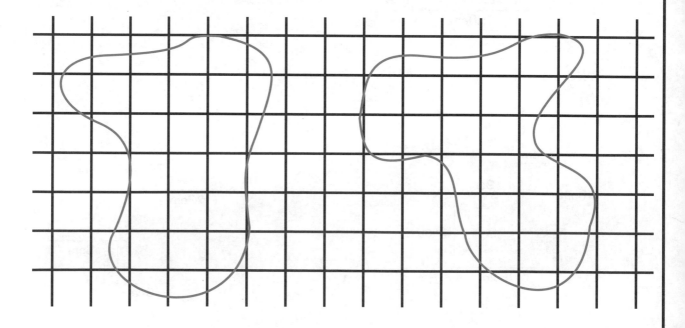

Area .... km²   Area .... km²

Add up all the whole squares first — then round off all the bits of squares to the nearest half, then add them all on afterwards.

SECTION TWO — SHAPES

2.3 Questions on Areas of Rectangles

For rectangles and squares, working out the area is a piece of pie — it's just **LENGTH TIMES WIDTH.** Nowt more to it.

AREA = LENGTH × WIDTH

Q1 Calculate the areas of the following :

a) Length = 10 cm, Width = 4 cm,   Area = ... × ..... = ... cm².

b) Length = 55 cm, Width = 19 cm, Area = ... cm².

c) Length = 12 m,   Width = 7 m,   Area = ... m².

d) Length = 155 m, Width = 28 m,   Area = ... m².

e) Length = 3.7 km, Width = 1.5 km, Area = ... km².

Q2 Measure the Lengths and Widths of each of these rectangles, then calculate the Area.

a)

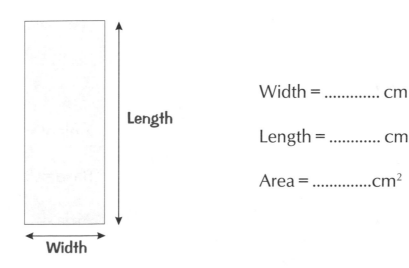

Width = ... cm

Length = ... cm

Area = ...cm²

b)

Width = ... cm

Length = ... cm

Area = ...cm²

2.4 *Questions on Areas of Triangles*

Triangles aren't much different, but <u>remember</u> to TIMES BY THE ½.

Area = ½ (Height × Base)

Q1   Calculate the areas of the following :

    a) Base = 12 cm, Height = 9 cm,   Area = $\frac{1}{2}$ (...... × ......) = ... cm².

    b) Base = 5 cm, Height = 3 cm,   Area = ... cm².

    c) Base = 25 m, Height = 7 m,   Area = ... m².

    d) Base = 1.6 m, Height = 6.4 m,   Area = ... m².

    e) Base = 700 cm, Height = 350 cm,   Area = ... cm².

Q2   Measure the Base and Height of each of these triangles, then calculate the Area.

    a)

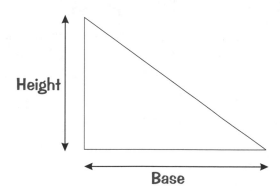

Height = ... cm

Base = ... cm

Area = ...cm²

    b)

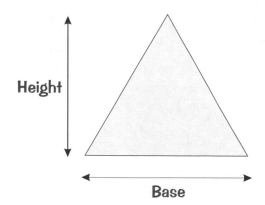

Height = ... cm

Base = ... cm

Area = ...cm²

SECTION TWO — SHAPES

2.5 Questions on Composite Areas

Calculate the areas of these composite shapes...

Q1

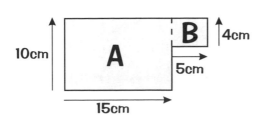

Shape A: length =........ width = ...
Area = ... × .... = ...cm²
Shape B: length = ... width = ...
Area = ... × .... = ...cm²
Total (area A + area B) = ... + ... = ...cm²

Q2 Shape A (rectangle): .... × .... = ...cm²
Shape B (triangle): ½ (base x height)
Base = ...cm , Height = ...cm
Area = ½ (........ × ....) = ...cm²
Total Area = ... + ... = ...cm²

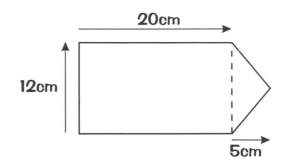

Q3

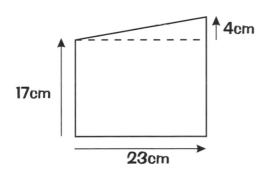

Shape A (rectangle): .... × .... = ...cm²
Shape B (triangle): ½ (....... × ....) = ...cm²

Total Area = ... + ... = ...cm²

Q4 Draw on the dotted line to divide this shape.
Shape A (rectangle): .... × .... = ...m²
Shape B (triangle): ½(....... × ....)= ....m²
Total Area = ... + ... = ...m²

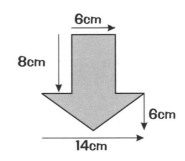

Q5

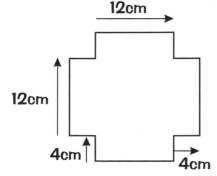

Draw on the dotted lines.
Shape A (rectangle): .... × .... = ...m²
Shape B (square): .... × .... = ...m²
Total Area = (4 × Shape A) + Shape B =.........m²

Bit more tricky, these... but look — they're all just rectangles and triangles.
Work out each bit separately, then add the areas together — easy.

2.6   Questions on The Circle

Don't worry about that π bit — it just stands for the number 3.14159... and then it's rounded off to **3** or **3.14**, to make it a bit easier for you to use.

Q1   Draw a circle with radius 3cm. On your circle label the circumference, a radius and a diameter.

Q2   Calculate the circumference of these circles.
Take π to be 3.14.

    a)   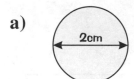   Circumference = π × diameter = ...

    b) Circumference = π × diameter = ...

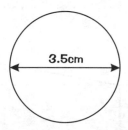

    c)

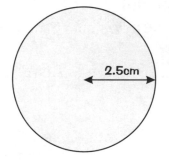

Remember to work out the diameter first.

Diameter = radius × 2 = ...

Circumference = ...

    d) A circle radius 3.5 cm. ....

Q3   A coin has a diameter of 1.7cm. What is its circumference?

....

Q4   A plant is in a pot. The radius of the top of the pot is 4.5cm.
Calculate the circumference of the pot.

....

Q5   The diameter of the wheels on Joe's
bike is 0.6m. On the way to school the
front wheel rotates 600 times. How far
does Joe live from school?

....

2.6   *Questions on The Circle*

Q6   Calculate the area of each circle.

a)

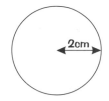

Area = π × radius² = ...

b) A circle of radius 9cm.   Area = π × radius² = ...

c)

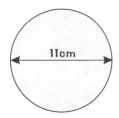

You must find the radius first.

Radius = diameter / 2 = ...

Area = ...

d) A circle of diameter 28cm.   Radius = ...

           Area = ...

Q7   Find the area of one face of a 10p coin, radius 1.2cm.

           Area = ...

Q8   A circular table has a diameter of 50cm.   Find the area of the table.

           Area = ...

Q9   This circular pond has a circular path around it. The radius of the pond is 72m and the path is 2m wide.

     What is the area of the pond? ....

     What is the area of the path? ....

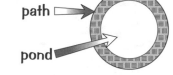

Q10   What is the area of this semicircular rug?

     Area = ...

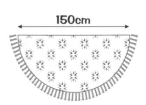

Q11   The circumference of a circle is 195cm.
       Calculate its diameter.

           Diameter = ...

Do exactly the same with these areas as you did before — add or take away the areas of individual bits. And don't forget a semicircle is half a circle...

2.7   *Questions on Solids and Nets*

 There's usually more than one possible net for each shape, so don't worry if you get a couple of answers — as long as yours works, you'll get the marks.

Q1   Which of the following nets would make a cube?

a)   **b)**   **c)**

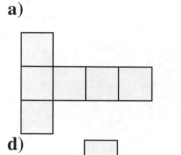   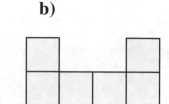   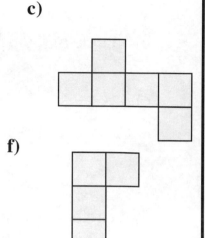

d)   **e)**   **f)**

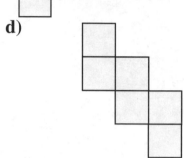   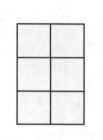

Q2   Below is a sketch of a cuboid and its net. The net is drawn to scale but not finished: it needs two more faces. Draw them in the correct position.

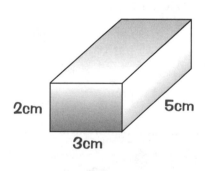   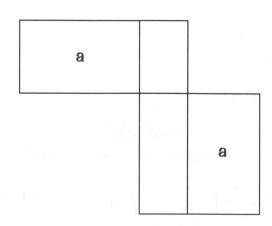

Q3   What is the total surface area of the cuboid in **Q2**?

Area = ...

2.7 Questions on Solids and Nets

Q4 On a 6-sided dice, opposite numbers should add up to 7. Fill in the rest of the dots on this net:

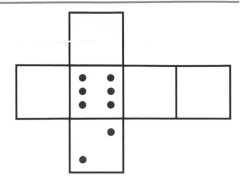

Q5 This unfinished isometric drawing shows a cuboid with dimensions 1 cm by 4 cm by 3 cm.

a) Complete the isometric drawing of the cuboid.

b) Draw the front elevation, side elevation and plan of the cuboid in the space below. Make sure your drawings are to scale.

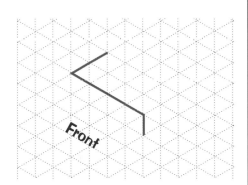

Q6 Draw an accurate net for each of the following solid shapes. (Use a spare page in this workbook or a separate piece of paper.)

a)

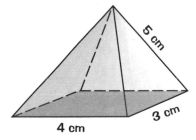

b)

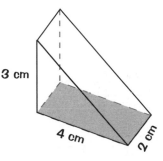

Q7 This net will make a common, mathematical solid. Name the solid:

....

You've got to think about folding the net up to make the shape — if you're struggling, the best thing to do is practise your origami skills...

2.8 Questions on Volume

Q1 Each shape has been made from centimetre cubes. The volume of a centimetre cube is 1 cubic cm. How many cubes are there in each shape? What is the volume of each shape in cubic cm?

a)

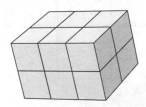

There are ...... cubes.
The volume is
...... cubic cm.

b)

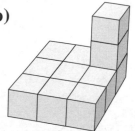

There are ...... cubes.
The volume is
...... cubic cm.

c)

There are ...... cubes.
The volume is
...... cubic cm.

d)

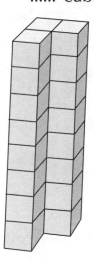

There are ...... cubes.
The volume is
...... cubic cm.

e)

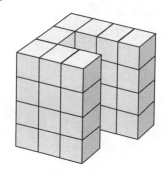

There are ...... cubes.
The volume is
...... cubic cm.

f)

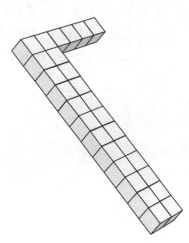

There are ...... cubes.
The volume is
...... cubic cm.

g)

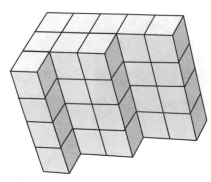

There are ...... cubes.
The volume is ...... cubic cm.

h)

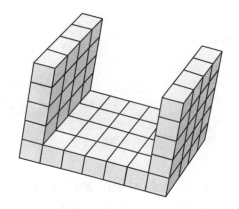

There are ...... cubes.
The volume is ...... cubic cm.

You simply add up the cubes... but make sure you don't miss any — remember that there are some rows at the back too.

SECTION TWO — SHAPES

2.8 Questions on Volume

Volumes of cubes and cuboids are nearly as easy as areas of squares and rectangles — you've only got an extra side to multiply by.

Q2 A match box measures 7cm by 4cm by 5cm. What is its volume?

....

Q3 A cereal box measures 30cm by 6cm by 15cm. What is its volume?

....

Q4 A room is 2.5m tall, 8m long and 5m wide. What is its volume?

....

Q5 Which holds more, a box measuring 12cm by 5cm by 8cm or a box measuring 10cm by 6cm by 9cm?

....

Q6 A video casette case measures 20cm by 3cm by 10cm. What is its volume?

....

Q7 An ice cube measues 2cm by 2cm by 2cm. What is its volume? ....
Is there enough room in a container measuring 8cm by 12cm by 10cm for 100 ice cubes?

....

Q8 What is the volume of a cube of side:

   a) 5cm ....

   b) 9cm ....

   c) 15cm .... ?

Q9 A box measures 9cm by 5cm by 8cm. What is its volume? ....
What is the volume of a box twice as long, twice as wide, and twice as tall?

....

2.8   *Questions on Volume*

 This looks a bit familiar... one tricky shape made up of a few easier shapes.
(You know what to do with this — individual volumes first, then add together)

Q1 Calculate the volume of these podiums...

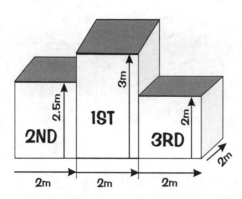

1st Place = length × width × height
= ..... × ...... × ...... = ...m³

2nd Place = ... × ..... × ..... = ...m³

3rd Place = ... × ..... × ..... = ...m³

Total = ... + ... + ... = ...m³

Q2 a) Which of these blocks of flats has the biggest volume?

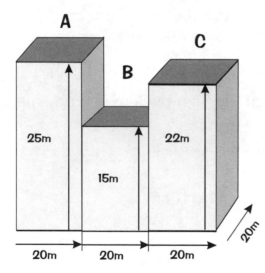

Block A = ... × ...... × ...... = ...m³

Block B = ... × ...... × ...... = ...m³

Block C = ... × ...... × ...... = ...m³

Biggest Volume = ...

b) Total Volume = ... m³

Q3 How much water would this tank hold when full to the brim?

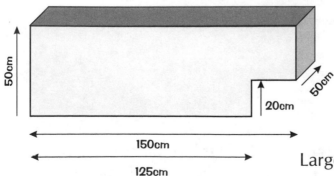

Large Part = ... × ...... × ...... = ...cm³

Small Part = ... × ...... × ...... = ...cm³

Total = ...cm³

2.9 Questions on Vertices, Faces and Edges

Q1 Fill in the boxes in the table.

	Name of SHAPE	number of FACES	number of EDGES	number of VERTICES

Q2 What are the names of these shapes?

a)

b)

c)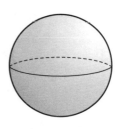

....

....

....

Name that shape... they're really keen on putting these in the Exam — gonna have to get learning those shape names, aren't you...

2.10 Questions on Symmetry

 You've got a line of symmetry if you can draw a line across the picture so both sides fold exactly together. Have a go yourself — it'll be lots of fun...

Q1 These shapes have only one line of symmetry. Draw the line of symmetry using a dotted line.

a)

b)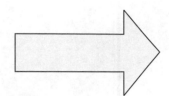

c)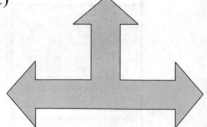

Q2 These shapes have more than one line of symmetry. Draw the lines of symmetry using dotted lines.

a)

b)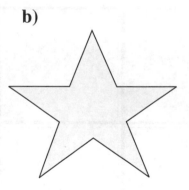

c)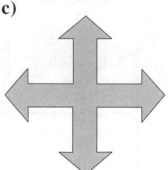

Q3 Some of the letters of the alphabet have lines of symmetry. Draw the lines of symmetry using dotted lines.

A B C D E F G H I J K L M

N O P Q R S T U V W X Y Z

Q4 Some of the counting numbers have a line of symmetry. Draw the line of symmetry using a dotted line.

1 2 3 4 5 6 7 8 9

2.10 Questions on Symmetry

Q5 Draw the mirror line.

a)

b)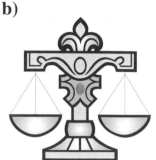

c)

Q6 Reflect the word in the X axis so that the X axis is the line of symmetry. Reflect the word in the Y axis so that the Y axis is the line of symmetry.

a)

Y axis

COMPUTER

X axis

b)

Y axis

MOUSE

Q7 The dotted line represents a line of reflectional symmetry. Draw the reflected pictures and label the points: A′ B′ C′ D′

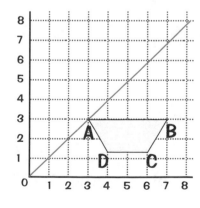

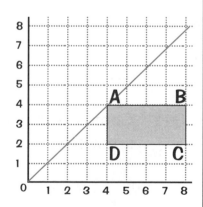

a) What are the coordinates of: A′( ) B′( ) C′( ) D′( )
b) What are the coordinates of: A′( ) B′( ) C′( ) D′( )
c) What are the coordinates of: A′( ) B′( ) C′( ) D′( )

MIRROR LINE = LINE OF SYMMETRY = AXIS OF SYMMETRY
— they'll use all of these words, but they mean the same thing.

SECTION TWO — SHAPES

2.10   Questions on Symmetry

 You can work out the rotational symmetry by sticking your pen in the middle of the shape and spinning your book round — how many times does the shape look the same before the book's back the right way up?

Q8   Write down the order of rotational symmetry of each of the following shapes:

a)

square

b)

rectangle

c)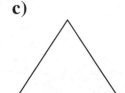

equilateral triangle

d)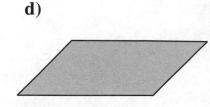

parallelogram

Q9   What is the order of rotational symmetry of the following capital letters?

e)   **f)**   **g)**   **h)**

N   T   S   C

Q10   Find the order of rotational symmetry for the following shapes:

i)    **j)**    **k)**    **l)**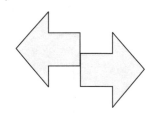

Q11   Complete the following diagrams so that they have rotational symmetry about centre C of the order stated:

m)   order 2   **n)**   order 4   **o)**   order 3

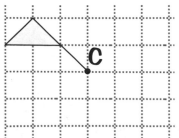   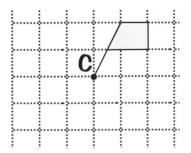   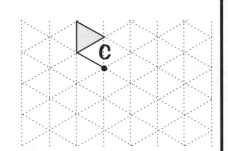

2.10   *Questions on Symmetry*

Q12   Which of the following shaded planes are planes of symmetry?

a)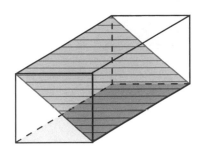

b)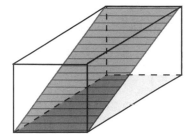

c)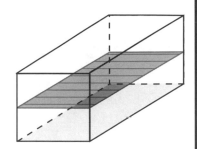

Q13   Give 6 examples of a plane of symmetry for a cube: (The first one is done for you)

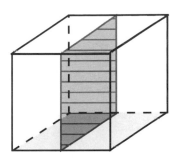

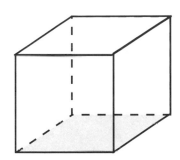

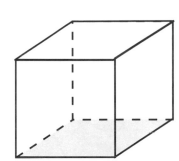

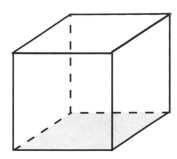

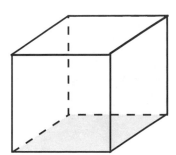

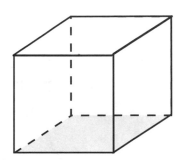

Planes of symmetry are a bit of a pain to draw, but if you shade them in a different colour from the rest of the object they'll be a lot easier to see.

2.11   *Questions on Tessellations*

Q1   A tessellation is a pattern made with identical 2-D shapes which fit together exactly leaving no gaps.

Continue these tessellations.

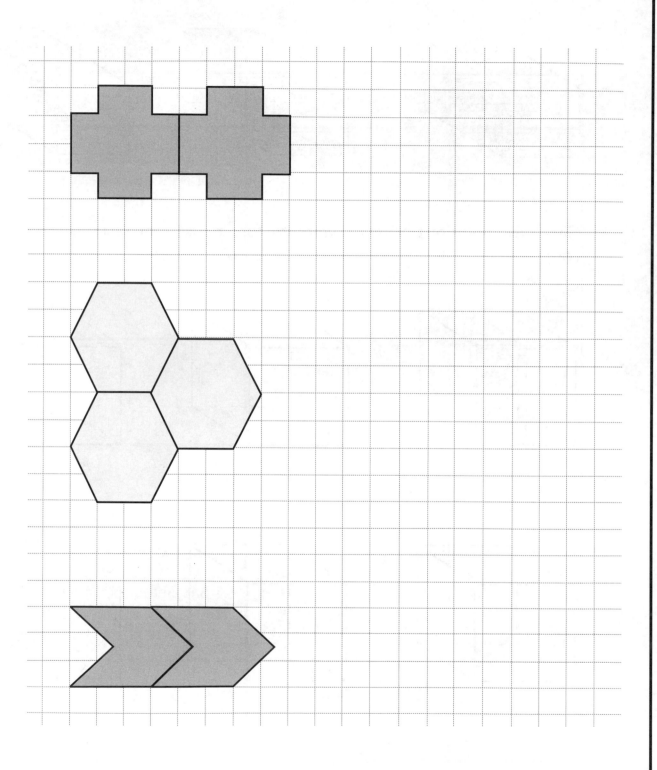

 Don't be put off by the word — a tessellation is just a fancy name for a jigsaw... where all the pieces are the same shape.

2.12   *Questions on Quadrilaterals*

Here's a few easy marks for you — all you've got to do is remember the shapes and a few facts about them... it's a waste of marks not to bother.

Q1   Fill in the blanks in the table.

NAME	DRAWING	DESCRIPTION
Square		Sides of equal length. Opposite sides parallel. Four right angles.
....		Opposite sides parallel and the same length. Four right angles.
....		Opposite sides are .... and Equal. Opposite angles are equal.
Trapezium		Only .... sides are parallel.
Rhombus		A parallelogram but with all sides ....
Kite		Two pairs of adjacent equal sides.

2.13   *Questions on Family Triangles*

Q1   Fill in the gaps in these sentences.

    a) An isosceles triangle has .... equal sides and .... equal angles.

    b) A triangle with all its sides equal and all its angles equal is called an ....
    triangle.

    c) A scalene triangle has .... equal sides and .... equal angles.

    d) A triangle with one right-angle is called a .... triangle.

Q2   By joining dots draw four different isosceles triangles, one in each box.

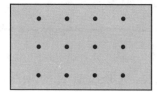 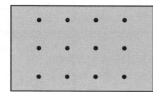 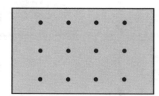 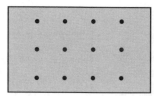

Q3   Using three different coloured pencils:
Find an equilateral triangle and shade it in.
Using a different colour, shade in two
different right angled triangles.
With your last colour shade in two different
scalene triangles.

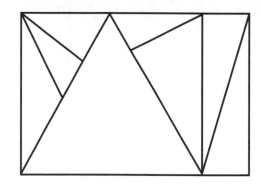

Q4   How many triangles are there in this diagram?

....

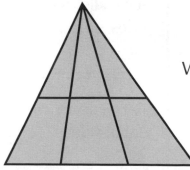

What sort of triangles are they?

....

There are only 4 types of triangles, so make sure you know them all — think
of all those nice fat juicy marks...

2.14   Questions on Regular Polygons

A polygon is a shape with lots of sides. Regular just means all the sides are
the same length and all the angles are the same. So a regular polygon is...

Q1   Can you name these regular shapes?

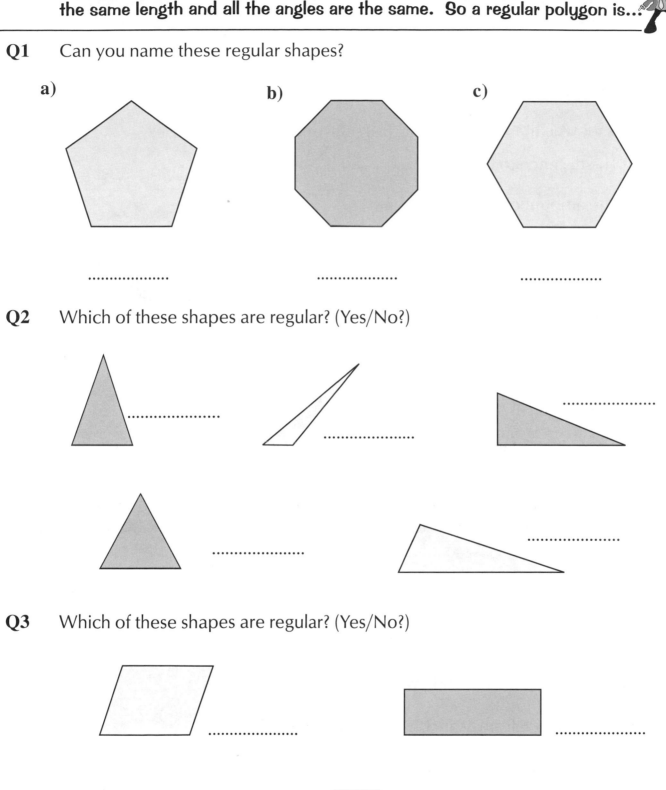

a)   b)   c)

....

Q2   Which of these shapes are regular? (Yes/No?)

Q3   Which of these shapes are regular? (Yes/No?)

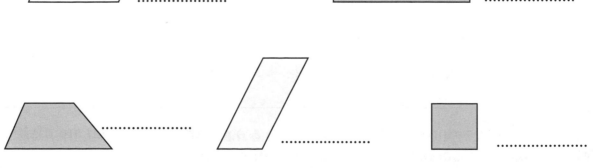

3.1   Questions on Metric and Imperial Units

| km | m | cm | mm | tonne | kg | g | mg | l | ml |

Q1   Which metric units from the box would you use to measure these in?

a) The length of your bedroom   ....

b) Your weight   ....

c) The distance to Paris   ....

d) The amount of water in the bath   ....

e) The weight of a packet of crisps   ....

f) The length of your finger   ....

g) The amount of medicine in a teaspoon   ....

h) The thickness of a coin   ....

i) The weight of a bus   ....

j) The weight of a butterfly   ....

| mile, | yard, | feet, | inch, | ton, | stone, | pound, | ounce, | gallon, | pint |

Q2   Which imperial unit goes in the gap? Choose from the box.

a) Paul's mother had a baby today. He weighs about 7 ....

b) A bottle of milk contains 1 ....

c) Alice is 15 years old. She weighs about 8 ....

d) The distance from Manchester to London is 203 ....

e) Jane's father is 6 .... tall.

f) This page is nearly 12 .... long.

You'll need to remember which units are imperial and which are metric — and make sure you know all the abbreviations for metric units.

3.1 Questions on Metric and Imperial Units

You'd better learn **ALL** these conversions —
you'll be well and truly scuppered without them.

APPROXIMATE CONVERSIONS
1kg = 2.2lbs   1gall = 4.5l   1in = 2.5cm
1 litre = 1.75pints   5 miles = 8km

Q3 The table shows the distances in miles between 4 towns in Scotland. Fill in the blank table with the equivalent distances in kilometres.

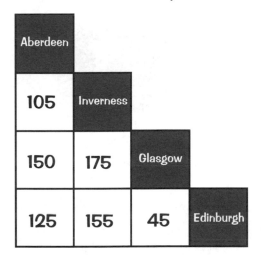

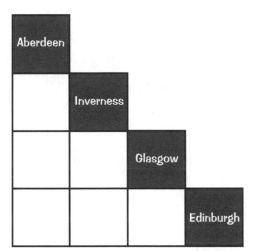

Q4 Change each of these weights from kilograms to pounds.

10kg = ... lbs   16kg = ... lbs   75kg = ... lbs

Change each of these capacities in gallons to litres.

5galls = ... l   14galls = ... l   40galls = ... l

Q5 Convert the measurements of the note book and pencil to centimetres.

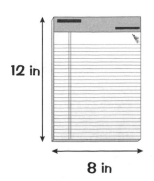

12in = ... cm

8in = ... cm

5in = ... cm

3.2   *Questions on Rounding Off*

 You always round off to the **NEAREST NUMBER**. It's a bit more tricky if it's exactly 1/2 **WAY** between 2 numbers — and then you just round **UP**.

Q1   Give these amounts to the nearest pound:

a) £4.29 ....   b) £16.78 ....   c) £12.06 ....

d) £7.52 ....   e) £0.93 ....   f) £14.50 ....

g) £7.49 ....   h) £0.28 ....

Q2   An average family has 2.3 children, how many children is this to the nearest whole number?

....

Q3   Round the following to the nearest whole number:

a) 2.9 ....   b) 26.8 ....   c) 2.24 ....

d) 11.11 ....   e) 6.347 ....   f) 43.5 ....

g) 9.99 ....   h) 0.41 ....

Q4   By the time she is 25 the average woman will have driven 4.72 cars. What is this to the nearest whole number?

....

Q5   Give these amounts to the nearest hour:

a) 2 hours 12 minutes ....   b) 36 minutes ....

c) 12 hours 12 minutes ....   d) 29 minutes ....

e) 100 minutes ....   f) 90 minutes ....

3.2   Questions on Rounding Off

Q6   Round off these numbers to the nearest 10:

     a) 23 = ...   **b)** 78 = ...   **c)** 65 = ...   **d)** 99 = ...

     e) 118 = ...   **f)** 243 = ...   **g)** 958 = ...   **h)** 1056 = ...

Q7   Round off these numbers to the nearest 100:

     a) 627 = ...   **b)** 791 = ...   **c)** 199 = ...   **d)** 450 = ...

     e) 1288 = ...   **f)** 3329 = ...   **g)** 2993 = ...

Q8   Round these off to the nearest 1000:

     a) 5200 = ...   **b)** 8860 = ...   **c)** 9870 = ...

Q9   Crowd sizes at sports events are often given exactly in newspapers. Round off these exact crowd sizes to the nearest 1000:

     a) 23324 = ...

     b) 36844 = ...

     c) 49752 = ...

Q10   The number of drawing pins in the box has been rounded to the nearest 10.

What is the least number of drawing pins in the box? ....

What is the greatest number?

....

Q11   The population of Whichtown is given as 1300 to the nearest 100. What is the smallest number the population could be?

....

What is the largest it could be?

....

When you round numbers off to the nearest unit, the ACTUAL measurement could be up to HALF A UNIT bigger or smaller...

3.3   *Questions on Conversion Factors*

Here are some more conversions that you've got to learn — these are even easier, so get memorising...

10 mm = 1 cm	1000 mg = 1 g
100 cm = 1 m	1000 g = 1 kg
1000 m = 1 km	1000 ml = 1 l

Q1   Complete this crossnumber using the conversion factors above.

ACROSS
1) 2cm to mm
2) 3500cm to m
4) 1.02m to cm
5) 0.5kg to g
6) 6.7km to m
7) 890 000g to kg
DOWN
1) 2.4m to cm
2) 3l to ml
3) 5.2kg to g
4) 1.57kg to g
6) 69 000ml to l

Q2   Fill in the gaps using the conversion factors:

20mm = ... cm	82mm = ... cm	.........mm = 6cm
142cm = ... m	.... cm = 2.5m	2550mm = ... m
9000m = ... km	3470m = ... km	.... m = 2km
3km = ... cm	.... mm = 3.4m	.... cm = 0.5km
6200mg = ... g	8550g = ... kg	2.3kg = ... g
12 000 000mg = ... kg	7.5kg = ... mg	
1.2l = ... ml	4400ml = ...l	6.75l = ... ml

3.4   *Questions on Decimal Places*

Q1   Round off these numbers to 1 decimal place (1 d.p.):

    a) 7.34 = ...   **b)** 8.47 = ...   **c)** 12.08 = ...   **d)** 28.03 = ...

    e) 9.35 = ...   **f)** 14.618 = ...   **g)** 30.409 = ...

Q2   Round off the following to 2 d.p.

    h) 17.363 = ...   **i)** 38.057 = ...   **j)** 0.735 = ...

    k) 5.99823 = ...   **l)** 4.297 = ...   **m)** 7.0409 = ...

Q3   Now round these to 3 d.p.

    n) 6.3534 = ...   **o)** 81.64471 = ...   **p)** 0.0075 = ...

    q) 53.26981 = ...   **r)** 754.39962 = ...

    s) 0.000486 = ...   **t)** 121.607593 = ...

Q3   Seven people have a meal in a restaurant. The total bill comes to £60. If they share the bill equally, how much should each of them pay? Round your answer to 2 d.p.

    ....

Q4   A 5m length of wood is cut into 12 equal pieces. How long is each of the pieces? Round your answer to 3 d.p.

    ....

Remember rounding to whole numbers... well this exactly the same — only slightly different.

3.5   Questions on Estimating

Q1   Estimate the answers to these questions...

For example: 12 × 21   <u>10 × 20 = <u>200</u></u>

a) 18 × 12   .... × .... = ...
       b) 23 × 21   .... × .... = ...

c) 57 × 46   .... × .... = ...
       d) 98 × 145   .... × .... = ...

e) 11 ÷ 4   .... ÷ .... = ...
       f) 22 ÷ 6   .... ÷ .... = ...

g) 97 ÷ 9   .... ÷ .... = ...
       h) 147 ÷ 14   .... ÷ .... = ...

i) 195 × 205   .... × .... = ...
       j) 545 × 301   .... × .... = ...

k) 901 ÷ 33   .... ÷ .... = ...
       l) 1207 ÷ 598   .... ÷ .... = ...

Q2   Write in the estimates that give the answer shown...

For example: 101 × 96   <u>100 × 100 = <u>10000</u></u>

a) 34 × 19   .... × .... = 600
       b) 27 × 32   .... × .... = 900

c) 67 × 89   .... × .... = 6300
       d) 99 × 9   .... × .... = 1000

e) 56 ÷ 11   .... ÷ .... = 6
       f) 119 ÷ 17   .... ÷ .... = 6

g) 182 ÷ 62   .... ÷ .... = 3
       h) 317 ÷ 81   .... ÷ .... = 4

Q3   Andy earns £12,404 a year. Bob earns £58,975 a year. Chris earns £81,006 a year.

a) Estimate how much Andy will earn over 3 years. £.............

b) Estimate how many years Andy will have to work to earn as much as Chris does in one year. ....

c) Estimate how much Bob earns per month. £.............

Round off to NICE EASY CONVENIENT NUMBERS, then use them to do the sum. Easy peas.

3.5   *Questions on Estimating*

An estimate isn't just a wild guess — you've usually got to do SOME work.

Q4

The ranger is almost 2m tall. Estimate the height of the giraffe in metres.

....

Q5   Estimate the following lengths then measure them to see how far out you were:

OBJECT	ESTIMATE	ACTUAL LENGTH
a) Length of your pen or pencil	....	....
b) Width of your thumb nail	....	....
c) Height of this page	....	....
d) Height of the room you are in	....	....

Q6   The distance from A to B is 50km. Estimate the distance from A to C then measure and work it out.

AC = ... km

• A   B •

• C

58

3.6 Questions on Conversion Graphs

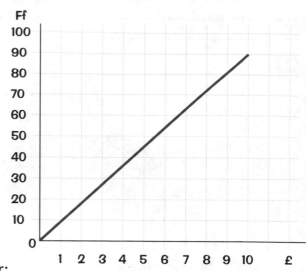

Q1 This graph shows the exchange rate between British Pounds (£) and French Francs (Ff). How many French Francs would I get for:

a) £10 ....

b) £4 ....

c) £40 ....

Q2 How many British Pounds would I get for:

a) Ff 70 ....

b) Ff 20 ....

c) Ff 1400 ....

Q3 This graph can be used to convert the distance (miles) travelled in a taxi to the fare payable (£). How much will the fare be if you travel:

a) 2 miles ....

b) 5 miles ....

c) 10 miles ....

How far would you travel if you paid:

d) £5 ....

e) £11 ....

f) £14 ....

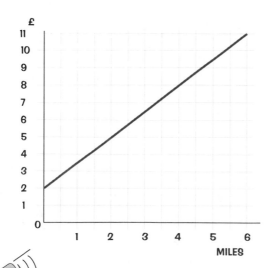

Remember conversion graphs can be read 2 ways — you can convert from one thing to the other and back again.

SECTION THREE — MORE NUMBERS

3.6   Questions on Conversion Graphs

When you've got to draw your own conversion graphs, your best bet is to work out a few different values, and mark them on the graph first.

Q5   80km is roughly equal to 50 miles, use this information to draw a conversion graph on the grid. Use the graph to estimate the number of miles equal to:

a) 20 km ....

b) 70 km ....

c) 90 km ....

Q6   How many km are equal to:

a) 40 miles ....

b) 10 miles ....

c) 30 miles ....

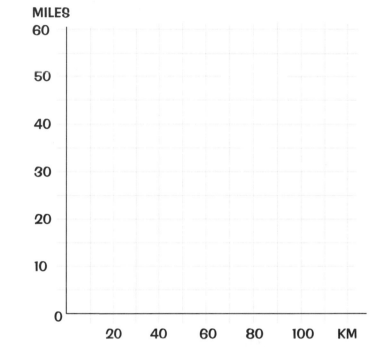

Q7   To hire a bouncy castle you need to first pay a minimum charge of £20.

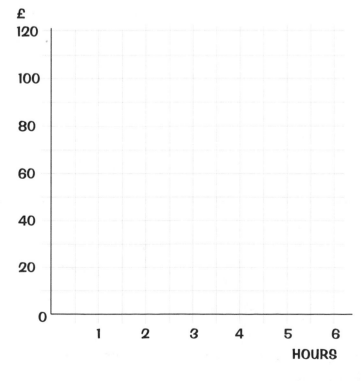

The cost then increases according to the length of time, for example to hire it for 5 hours costs a total of £100. Use this information to draw a conversion graph on the grid. Use the graph to estimate the cost for hiring for:

a) 3 hours ....

b) 6 hours ....

c) 1 hour ....

3.7 Questions on Recognising Fractions

 Yeah, OK, we know a **FRACTION**'s a **PART** of a **WHOLE** — but you've gotta learn the names of the top (**NUMERATOR**) and bottom (**DENOMINATOR**), too.

 This circle is divided into two equal parts — each bit is 1/2 of the whole.

Q1 What fraction is shaded in each of the following pictures?

a)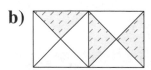

....

b)

....

c)

....

d)

....

e)

....

f)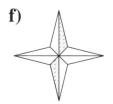

....

g)

....

Q2 Shade the diagram to show the following fractions.

a) $\frac{3}{8}$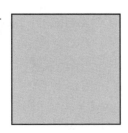

b) $\frac{2}{5}$

c) $\frac{7}{10}$

Q3 There are 32 boxes of shoes stacked on a shelf.

If five eighths are sold, can you show on the diagram how many are left?

3.8   Questions on Equivalent Fractions

Q1   Shade in the correct number of sections to make these diagrams equivalent...

$\frac{1}{4} =$

$\frac{1}{3} =$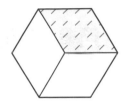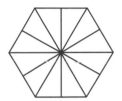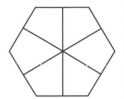

Q2   Write in the missing numbers to make these fractions equivalent.

For example 1/2 = 7/14

a) 1/4 = 4/........   b) 3/4 = 9/........   c) 1/3 = ... /6

d) 2/3 = 8/........   e) 2/7 = 6/........   f) 6/18 = 1 /........

g) 8/16 = ... /2   h) 24/32 = 3/........   i) 10/60 = 5/........

Q3   Write in the missing numbers to make each list equivalent.

a) 1/2 = 2/...... = ... /6 = ... /8 = 5/10 = 25/...... = ... /70 = ... /100

b) 200/300 = 100/ ...... = ... / 15 = 40/ ...... = 120/180 = ... / 9 = .../3

c) 7/10 = 14/ ...... = ... / 30 = 210/ ...... = 49/ ...... = ... /20

d) 19/20 = ... /80 = 38/ ...... = 57/ ...... = ... /100 = ... /1000

e) 500/600 = 250/ ...... = 50 / ...... = ... / 150 = 1000/ ....

Q4   Which is bigger?

a) 1/5 or 2/10 ....   b) 3/7 or 6/21 ....

c) 10/15 or 4/6 ....   d) 1/3 or 33/100 ....

**To make an EQUIVALENT fraction, you've got to multiply the TOP
(numerator) and BOTTOM (denominator) by the SAME THING.**

3.9   *Questions on Improper Fractions*

Here is a proper fraction $\frac{3}{4}$.

Here is an improper fraction $\frac{14}{3}$.   Yes — it's Top-Heavy.

Here is a mixed number $2\frac{3}{5}$.

To enter this into your calculator you would key in 2 $\boxed{a^b/_c}$ 3 $\boxed{a^b/_c}$ 5 and you would see $\boxed{2 \llcorner 3 \llcorner 5}$ or $\boxed{2 \ulcorner 3 \ulcorner 5}$.

Change these top-heavy fractions to mixed numbers:

a) $\frac{3}{2} =$ **b)** $\frac{7}{4} =$ **c)** $\frac{8}{3} =$

d) $\frac{17}{5} =$ **e)** $\frac{18}{6} =$ **f)** $\frac{25}{11} =$

Change these mixed numbers to top-heavy fractions:

g) $2\frac{1}{2} =$ **h)** $3\frac{1}{3} =$ **i)** $1\frac{3}{5} =$

j) $4\frac{2}{3} =$ **k)** $3\frac{5}{8} =$ **l)** $10\frac{2}{7} =$

Try these fraction additions:

m) $\frac{1}{5} + \frac{2}{5} =$ **n)** $\frac{3}{10} + \frac{4}{10} =$

o) $\frac{5}{8} + \frac{7}{8} =$ **p)** $\frac{1}{2} + \frac{1}{8} =$

 Make sure you know how to use the fraction button on your calculator — it'll save you time if you can use it in the Exam.

3.10 Questions on Fractions of Quantities

Q1 Write down the fractions of the following quantities...

a) Half of 12 = ...   b) Quarter of 24 = ...   c) Third of 30 = ...

d) 1/4 of 44 = ...   e) 3/4 of 60 = ...   f) 2/3 of 6 = ...

Q2 Calculate the fraction of the following ...

eg. 1/3 of 18 = 18 ÷ 3 = <u>6</u>

a) 1/8 of 32 = ... ÷ 8 = ...   b) 1/10 of 50 = ... ÷ 10 = ...

c) 1/12 of 144 = ... ÷ .... = ...   d) 1/25 of 75 = ... ÷ .... = ...

e) 1/30 of 180 = ... ÷ .... = ...   f) 1/27 of 540 = ... ÷ .... = ...

Q3 Calculate the fractions of the following...

eg 2/5 of 50   50 ÷ 5 = 10   2 × 10 = <u>20</u>

a) 2/3 of 60   60 ÷ 3 = ...   2 × .... = ...

b) 4/5 of 25   25 ÷ .... = ...   4 × .... = ...

c) 7/9 of 63   .... ÷ .... = ... × .... = ...

d) 3/10 of 100   .... ÷ .... = ... × .... = ...

e) 12/19 of 760   .... ÷ .... = ... × .... = ...

f) 6/9 of £1.80   .... ÷ .... = ... × .... = £....... or ....p

g) 10/18 of £9.00   .... ÷ .... = ... × .... = £..........

h) 2/3 of one day (24 hours)   .... ÷ .... = ... × .... = ...hours

i) 5/6 of one year (12 months)   .... ÷ .... = ... × .... = ...months

j) 2/5 of one kilogram (1000 grams)   .... = ...g

DIVIDE BY THE BOTTOM, TIMES BY THE TOP — and suddenly fractions don't seem so bad...

3.11   Questions on Fractions, Decimals, %

Q1   Change these fractions to decimals:

a) $\frac{1}{2}$ ....   b) $\frac{3}{4}$ ....   c) $\frac{7}{10}$ ....   d) $\frac{19}{20}$ ....

e) $\frac{1}{100}$ ....   f) $\frac{3}{8}$ ....   g) $\frac{2}{1000}$ ....   h) $\frac{1}{3}$ ....

Q2   Change these fractions to percentages:

a) $\frac{1}{4}$ ....   b) $\frac{3}{10}$ ....   c) $\frac{4}{5}$ ....   d) $\frac{12}{25}$ ....

e) $\frac{8}{100}$ ....   f) $\frac{2}{40}$ ....   g) $\frac{7}{8}$ ....   h) $\frac{11}{30}$ ....

Q3   Change these decimals to percentages:

a) 0.62 ....   b) 0.74 ....   c) 0.4 ....   d) 0.9 ....

e) 0.07 ....   f) 0.02 ....   g) 0.125 ....   h) 0.987 ....

Q4   Change these decimals to fractions (in their lowest terms if possible):

a) 0.5 ....   b) 0.8 ....   c) 0.19 ....   d) 0.25 ....

e) 0.64 ....   f) 0.06 ....   g) 0.125 ....   h) 0.075 ....

Q5   Change these percentages to fractions (in their lowest terms if possible):

a) 75% ....   b) 60% ....   c) 15% ....   d) 53% ....

Q6   Change these percentage to decimals:

a) 25% ....   b) 49% ....   c) 3% ....   d) 30% ....

A FRACTION IS A DECIMAL IS A PERCENTAGE — they're all just different ways of saying "a bit of" something.

3.12   *Questions on Percentages*

Only use the % button on your calculator when you're sure you know what it does — or things will just go pear-shaped.

Q1   Try these <u>without</u> a calculator:

a) 50% of £12 =...........   **b)** 25% of £20 =...........

c) 10% of £50 =...........   **d)** 5% of £50 =...........

e) 30% of £50 =...........   **f)** 75% of £80 =...........

g) 10% of 90cm =...........   **h)** 10% of 4.39kg =...........

Now you can use a calculator:

i) 8% of £16 =...........

j) 15% of £200 =...........

k) 12% of 50 litres =...........

l) 85% of 740kg =...........

m) 40% of 40 minutes =...........

n) 17½% of £180 =...........

A school has 750 pupils.

o) If 56% of the pupils are boys, what percentage is girls? ....

p) How many boys are there in the school? ....

q) One day, 6% of the pupils were absent. How many pupils was this? ....

r) 54% of the pupils have a school lunch, 38% bring sandwiches and the rest go home for lunch. How many pupils go home for lunch?

....

3.12   *Questions on Percentages*

1% is just 1 out of 100 — that's all there is to it... and it's worth learning — someday you'll be interested in working out what a 7% pay rise gives you...

Q2   Car workers at the Fiort plant are given a pay rise. Each grade of worker gets a different percentage (%) increase.

   a) Complete the table to show what each worker will now earn.

	Current Pay	% Rise	Extra Pay	New Pay
Manager	£35,000	7%		
Skilled	£24,000	5%		
Unskilled	£18,000	3%		

At another company things are going badly. Sales have fallen and the workers are asked to take a pay cut of 8% to save their jobs.

   b) Complete the table to show how much money each worker will lose.

	Current Pay	% Cut	Pay cut in £
Manager	£35,000	8%	
Skilled	£24,000	8%	
Unskilled	£18,000	8%	

Q3   VAT (value added tax) is charged at a rate of 17.5% on many goods and services. Complete the table showing how much VAT has to be paid and the new price of each article.

Article	Basic price	VAT at 17.5%	Price + VAT
Tin of paint	£6.75		
Paint brush	£3.60		
Sand paper	£1.55		

Q4   A company claims that its insulation material will cut home heating bills by 40%. The Coalman family currently pay £480 per year for heating their house. How much will they save each year if they buy this new insulation material?

Savings made each year ....

3.12　　　　*Questions on Percentages*

Q5　　Express each of the following as a percentage. Round off if necessary...

a) £6 of £12 = ...

b) £4 of £16 = ...

c) 600kg of 750kg = ...

d) 6 hours of one day = ...

e) 1 month of a year = ...

f) 25m of 65m = ...

Q6　　Calculate the percentage saving of the following:

eg. Trainers: Was £72 Now £56 Saved .... of £72 = ...%

a) Jeans: Was £45 Now £35 Saved .... of £45 = ...%

b) CD: Was £14.99 Now £12.99 Saved .... of .... = ...%

c) Shirt: Was £27.50 Now £22.75 Saved .... of .... = ...%

d) TV: Was £695 Now £435 Saved .... = ...%

e) Microwave: Was £132 Now £99 Saved .... = ...%

Q7 a) In their first game of the season, Nowcastle had 24,567 fans watching the game. By the final game there were 32,741 fans watching. What is the percentage increase in the number of fans? ....%

b) If Tim Hangman won 3 out of the 5 sets in the Wimbledon Men's Final, what percentage of the sets did he not win? ....%

c) Jeff went on a diet. At the start he weighed 92kg, after one month he weighed 84kg. What is his percentage weight loss? ....%

d) Of Ibiza's 25,000 tourists last Summer, 23750 were between 16 and 30 years old. What percentage of the tourists were not in this age group? ....%

Divide the new amount by the old, then × 100... or if you've been practising on your calc, you'll know you can just press the % button for the 2nd bit...

4.1   Questions on Mode

Remember the GOLDEN RULE — put things IN ORDER OF SIZE first.

The MODE or MODAL value is the one that occurs most often in a set of data.

Q1   Find the MODE for each of these sets of data.

     a) 3,   5,   8,   6,   3,   7,   3,   5,   3,   9,   Mode is ....

     b) 52,   26,   13,   52,   31,   12,   26,   13,   52,   87,   41   Mode is ....

Q2   The temperature in °C on 10 Summer days in England was:

         25,   18,   23,   19,   23,   24,   23,   18,   20,   19

     What was the Modal temperature?   Modal temperature ....°C.

Q3   The time it takes twenty pupils in a class to get to school each day in minutes is:

       18,   24,   12,   28,   17,   34,   17,   17,   28,   12

       23,   24,   17,   34,   19,   32,   15,   31,   17,   9

       17,   32,   15,   17,   21,   29,   34,   17,   12,   17

     What is the modal time?   Modal time..............mins.

Q4   The first thirty entrants in a competition were given an envelope with a sum of money in it. The amounts were:

       £5,   £10,   £5,   £1,   £20,   £20,   £10,   £5,   £10,   £20

       £10,   £20,   £10,   £5,   £10,   £5,   £10,   £10,   £5,   £20

       £1,   £5,   £10,   £5,   £20,   £1,   £20,   £10,   £5,   £10

     What was the modal amount ?   Modal amount £...........

Q5   A domino is selected twenty times and the dots added. The results are:

     What is the modal score?   Modal Score ....

4.2   *Questions on Median*

The Median is the middle value when the data has been put in order of size.

Q1   Find the median for these sets of data.

     a) 3,   6,   7,   12,   2,   5,   4,   2,   9

     ....   Median is ....

     b) 14,   5,   21,   7,   19,   3,   12,   2,   5

     ....   Median is ....

Q2   These are the heights of fifteen 16 year olds.

         162cm   156cm   174cm   148cm   152cm
         139cm   167cm   134cm   157cm   163cm
         149cm   134cm   158cm   172cm   146cm

What is the median height? Put it in the shaded box.

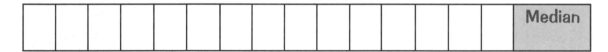

														Median

Q3   Some information about the seven members of the Harglobe basketball team is shown below.

Height	185cm	194cm	181cm	190cm	189cm	193cm	187cm
Weight	87kg	83kg	85kg	89kg	86kg	90kg	88kg
Age	24	26	21	25	23	24	29

Fill in the table below:

Median Height	Median Weight	Median Age

See, they're not too bad really — at least not when you use that Golden Rule.

4.2   Questions on Mean

 Yikes — MEAN questions... well, they're not as bad as everyone makes out. Remember to include zeros in your calculations — they still count.

The mean of a set of data is the total of the items ÷ the number of items

Q1   Find, <u>without</u> a calculator, the mean for each of these sets of data:

a) 5, 3, 7, 3, 2 =............   **b)** 18, 6, 12, 4 =...........

c) 7, 3, 9, 5, 3, 5, 4, 6, 2, 6 =...............   **d)** 5, 4, 0, 3, 0, 6 =...............

Q2   Now you can use a calculator to find the mean. If necessary, round your answers to 1 decimal place:

a) 13, 15, 11, 12, 16, 13, 11, 9 =..............

b) 16, 13, 2, 15, 0, 9 =..............

c) 80, 70, 80, 50, 60, 70, 90, 60, 50, 70, 70 =..............

Q3   Emily measured the heights of her five friends. The measurements in metres were:

1.65, 1.66, 1.70, 1.72, 1.67

What was the mean height of her friends?

....

Q4 a) Stephen scored a mean mark of 64 in four Maths tests. What was his total marks for all four tests?

....

b) When he did the next test, his mean mark went up to 66. What mark did he get in the fifth test?

....

4.3   *Questions on Range*

Q1   The number of goals scored by a hockey team over a period of 10 games is listed below.

0, 3, 2, 4, 1, 2, 3, 4, 1, 0.

What is the range in number of goals? ....

Q2   Sarah and her friends were measured and their heights were found to be:

1.52m, 1.61m 1.49m, 1.55m, 1.39m, 1.56m.

What is the range of the heights? ....

Q3   Here are the times 6 people took to do a Maths test:

1 hour 10 mins, 2 hours 10 mins, 1 hour 35 mins,
1 hours 55 min, 1 hour 18 mins, 2 hours 15 mins.

What is the range of these times? ....

Q4   The minimum temperatures over 5 nights were:

3°C, 2°C, –2°C, –1°C, 4°C.

What is the range of temperatures? ....

Q5   What is the range in the number of letters contained in each of the words in this question? ....

Q6   Some friends compared their pocket money. The least amount was £4 a week and the range was £16. How much was received by the person who got the most?

....

Q7   The weights of some people were surveyed. If the heaviest person was 70kg and the range of weights was 11kg, what was the weight of the lightest person?

....

Range is the difference between the biggest and smallest. Once you've found them, do a quick subtraction, then you can sit back and enjoy the scenery...

4.4   *Questions on Averages*

Geesh — as if it's not enough to make you work out all these boring averages, they want you to write stuff about them as well. Oh well, here goes nothing.

Q1   These are the mathematics marks for John and Mark.

John	65	83	58	79	75
Mark	72	70	81	67	70

Calculate the mean and range for each pupil. Who do you think is the better maths student? Why?

....

....

Q2   The shoe sizes in a class of girls are:

3 3 4 4 5 5 5 5 6 6 6 7 8

Calculate the mean, median and mode for the shoe sizes.

....

....

If you were a shoe shop manager, which average would be most useful to you, and why?

....

Q3   A house building company needs a bricklayer. This advert appears in the local newspaper.
The company employs the following people.

Position	Wage
Director	£600
Foreman	£260
Plasterer	£200
Bricklayer	£150
Bricklayer	£150

What is the median wage? ....

What is the mean wage? ....

Which gives the best idea of the average wage?

....

Is the advert fair? .... Why? ....

Write a fairer advert.

4.4   *Questions on Averages*

Q4 The number of absences for 20 pupils during the spring term were:

0 0 0 0 0 0 1 1 1 2 3 4 4 4 7 9 10 10 19 22

Work out the mean, median and modal number of absences.

....

....

....

If you were a local newspaper reporter wishing to show that the local school has a very poor attendance record, which average would you use and why?

....

If you were the headteacher writing a report for the parents of new pupils which average would you use and why?

....

Q5 On a large box of matches it says "Average contents 240 matches".
I counted the number of matches in ten boxes.
These are the results:

241   244   236   240   239   242   237   239   239 236

Is the label on the box correct? Use the mean, median and mode for the numbers of matches to explain your answer.

....

....

....

Q6 Jane has a Saturday job. She earns £2.20 an hour. She thinks that most of her friends earn more. Here is a list of how much an hour her friends are paid.

Lisa	£2.10	Scott	£3.00
Kate	£2.75	Kylie	£1.90
Helen	£2.51	Kirsty	£2.75
Ben	£2.75	Ruksana	£2.40

Work out the mean, median and mode for her friends' pay.

....

....

Jane's employer offers her a rise to £2.52 because she claims that this is the average hourly rate. Which average has her employer used?

....

Which average should Jane use to try to negotiate a higher pay rise?

....

The big trick with wordy questions is to IGNORE THE WAFFLE... you only need the numbers to work out the averages, so why worry about the rest of it...

SECTION FOUR — STATISTICS AND GRAPHS

4.5 Questions on Tally/Frequency Tables

 Tally charts are really simple — and they make sure you don't miss out any items. Worth doing, you've got to admit.

Q1 The results of a survey of 20 pupils, asked how many brothers and sisters they have, are:

0, 1, 0, 2, 4, 3, 3, 4, 2, 2, 2, 3, 2, 1, 2, 3, 0, 0, 2, 1

Fill in the tally chart below from this list of numbers, then total the frequency for each category in the end column.

NUMBER OF BROTHERS/SISTERS	TALLY	FREQUENCY
0		
1		
2		
3		
4		
	TOTAL	

Q2 The following list of shoe sizes needs to be put into a tally chart. Draw the chart and enter the numbers into the correct row. Finish by adding up the tally for each row and writing in the frequency.

5, 5, 3, 9, 10, 7, 6, 6, 5, 4, 9, 8, 9, 10, 5, 3, 4, 8, 7, 7, 4, 5, 8, 6, 6

	TOTAL	

4.5 Questions on Tally/Frequency Tables

Q3 At the British Motor Show 60 people were asked what type of car they preferred. Jeremy wrote down their replies using a simple letter code.

Saloon - S   Coupe - C   Hatchback - H   4x4 - F   MPV - M   Roadster - R

Here is the full list of replies.

H	S	R	S	S	R	M	F	S	S	R	R
M	H	S	H	R	H	M	S	F	S	M	S
R	R	H	H	H	S	M	S	S	R	H	H
H	H	R	R	S	S	M	M	R	H	M	H
H	S	R	F	F	R	F	S	M	S	H	F

Fill in the tally table and add up the frequency in each row. Draw the frequency graph of the results.

TYPE OF CAR	TALLY	FREQUENCY
Saloon		
Hatchback		
4 × 4		
MPV		
Roadster		

Q4 Last season Newcaster City played 32 matches.
The number of goals they scored in each match were recorded as shown.

2	4	3	5		1	0	0	1
1	0	3	2		1	1	1	0
4	2	1	2		1	3	2	0
0	2	3	1		1	1	0	4

Complete the tally chart and draw the frequency graph of the scores.

GOALS	TALLY	FREQUENCY
0		
1		
2		
3		
4		
5		

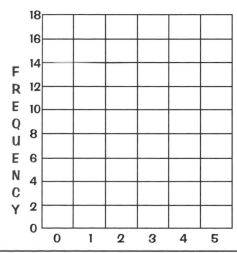

Go through the data in the <u>same order</u> as it's written and <u>cross off</u> each one as you put it in the list.

4.5 Questions on Tally/Frequency Tables

Q5 Here is a list of marks which 32 pupils gained in a History test:

65	78	49	72	38	59	63	44
55	50	60	73	66	54	42	72
33	52	45	63	65	51	70	68
84	61	42	58	54	64	75	63

Complete the tally table making sure you put each mark in the correct group. Then fill in the frequency column.

MARKS	TALLY	FREQUENCY
31-40		
41-50		
51-60		
61-70		
71-80		
81-90		
	TOTAL	

Q6 The weights in kilograms of 30 pupils were recorded as:

48	52	54	57	47	57	58	52	61	59
50	57	62	54	47	60	59	65	53	50
61	62	65	43	70	55	53	50	59	46

Complete the grouped frequency table choosing your own sensible group width:

Weight   Tally   Frequency

Don't forget to group your tally scores in 5's, using the 5 bar tally — ℍℍ.
Makes it all nice and tidy, doesn't it — and quicker to count.

SECTION FOUR — STATISTICS AND GRAPHS

4.5   *Questions on Tally/Frequency Tables*

Check your frequency column adds up to the number of items in your list.
If they don't you've gone wrong — which means you'll have to start again...

Q7   The frequency chart below shows the number of goals scored by Broughton Boys over 25 matches.

Number of Goals Scored	0	1	2	3	4
Frequency	6	7	10	0	2
Goals × Frequency					

a) What is the modal number of goals scored by the team? ....

b) Fill in the missing values in the Goals × Frequency row.

c) What is the total number of goals scored during the 25 matches? ....

d) Using your answer to **c)**, what is the mean number of goals scored by the team each game?...............................

Q8   The frequency table below shows the number of hours spent Christmas shopping by 100 people surveyed in a town centre.

Number of Hours	0	1	2	3	4	5	6	7	8
Frequency	1	9	10	10	11	27	9	15	8
Hours × Frequency									

a) What is the modal number of hours spent Christmas shopping? ....

b) Fill in the third row of the table.

c) What is the total amount of time spent Christmas shopping by the all the people surveyed?

....

d) What is the mean amount of time spent Christmas shopping by a person?

....

4.6 Questions on Tables, Charts and Graphs

Bar charts are a bit of a breeze — the bars' heights are just proportional to the frequencies they represent...

Q1 Complete this frequency table, and then draw a bar chart for the results.

TEST SCORE	TALLY	FREQUENCY
1 - 5	‖‖ ‖	6
6 - 10	‖‖ ‖‖‖	
11 - 15	‖‖‖	
16 - 20	‖‖‖‖	
21 - 25	‖‖‖	

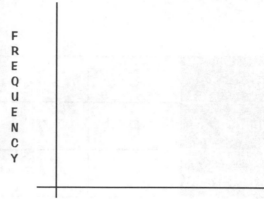

F R E Q U E N C Y

TEST SCORE

Q2 Complete this frequency table, and then draw a bar chart for the results.

AGE GROUP	TALLY	FREQUENCY
1 - 10	‖‖‖ ‖‖‖ ‖‖	
11 - 20	‖‖‖ ‖‖‖	
21 - 30	‖‖‖ ‖‖‖ ‖‖‖ ‖‖	
31 - 40	‖‖‖‖	
41 - 50	‖‖‖ ‖‖‖	

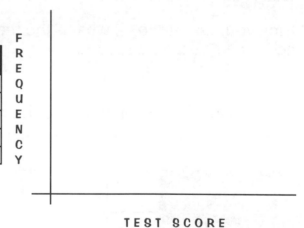

F R E Q U E N C Y

TEST SCORE

Q3 One hundred people were asked in a survey what colour eyes they had. Use this two-way table to answer the following questions.

a) How many people in the survey had green eyes? ....

b) How many women took part in the survey? ....

c) How many women had blue eyes? ....

d) How many men had brown eyes? ....

	Green eyes	Blue eyes	Brown eyes	Total
Male	15			48
Female	20		23	
Total		21		100

4.6 Questions on Tables, Charts and Graphs

Q4 The bar chart below shows the number of books that a group of pupils in Year 11 carried to school one day.

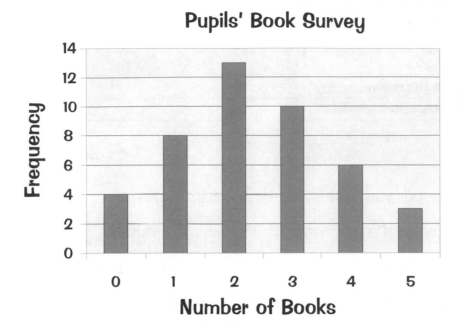

Pupils' Book Survey

a) How many pupils were in the survey? ....

b) What is the modal number of books brought to school? ....

c) How many books are there altogether in the bar chart?

   0 × 4 = 0 then 1 × 8 = 8 and continue...

   ....

   ....

d) Work out the mean number of books brought to school:

   The Total Number of Books in Survey ÷ The Number of Pupils Surveyed

   ....

   ....

   ....

All bar charts are basically the same... so remember the method for parts c) and d) — you can use this on **ANY BAR CHART** to work out the mean.

4.7   *Questions on Pictograms*

Q1   This pictogram shows the favourite drinks of a group of pupils.

Favourite Drinks	Number of Pupils
Lemonade	✧ ✧ ✧ ✧ ✧ ✧ ✧ ✧ ✧
Coke	✧ ✧ ✧ ✧ ✧ ✧ ✧ ✧ ✧ ✧ ✧
Tango	✧ ✧ ✧ ✧ ✧ ✧
Orange Squash	✧ ✧ ✧
Milk	✧

✧ Represents 2 pupils.

a) How many pupils were questioned?............... pupils.

b) How many pupils don't like fizzy drinks?.............. pupils.

c) 18 pupils liked lemonade best. How many more liked coke best?..............pupils.

d) Make one general comment about the information given................

Q2   During the last 5 years lots of people have been buying bicycles. The pictogram shows the number of people buying bicycles.

Year	Number of people buying bikes
1994	🚲 🚲
1995	🚲 🚲 🚲
1996	🚲 🚲 🚲 🚲 🚲
1997	🚲 🚲 🚲 🚲 🚲 🚲 🚲
1998	🚲 🚲 🚲 🚲 🚲 🚲 🚲 🚲

🚲 Represents 8,000 people

a) How many people bought bikes in the last 5 years?...............people

b) How many people bought bikes in the last 2 years?................people.

c) How many more people bought bikes in 1998 than in 1994?...............people.

d) Why do you think so many people are now buying bikes?.................

These are even better than bar charts... just count the piccies, times by whatever they stand for, then sit back and enjoy the ride.

4.7   *Questions on Pictograms*

Unless you're a budding artist, I'd stick with simple symbols... no point doing a work of art, then finding you've got another 20 still to do.

Q3   In Spain on the Costa del Sol the average daily hours of sunshine for the winter months are shown in the table.

MONTH	October	November	December	January	February	March
SUN HOURS	7	6	5	6	6	8

a) Complete the pictogram to represent the information. Use your own symbol.

Month	Hours of Sunshine
October	

.............Represents............... hours of sunshine.

b) What is the total number of sunshine hours during the Winter?................hours.

Q4   At school pupils were asked about their favourite TV soaps. The results are shown in the table.

TV SOAP	Eastenders	Emmerdale	Neighbours	Hollyoaks	Brookside	Coro. St
PUPILS	250	50	200	100	200	50

a) Complete the pictogram to represent the information. Use your own symbol.

TV Soap	Number of Pupils
Eastenders	
Emmerdale	
Neighbours	
Hollyoaks	
Brookside	
Coronation Street	

.... Represents .... pupils.

b) What was the total number of pupils questioned? .... pupils.

SECTION FOUR — STATISTICS AND GRAPHS

4.8 Questions on Bar Charts

Q1 Here is a horizontal bar chart showing the favourite colours of a class of pupils.

  a) How many like blue best? ....

  b) How many more people chose red than yellow? ....

  c) How many pupils took part in this survey? ....

  d) What fraction of the class prefer green? ....

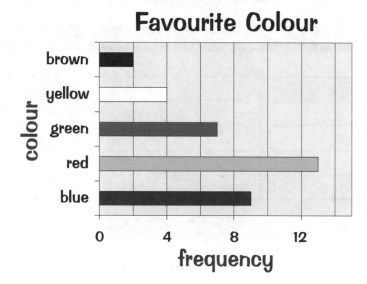

Favourite Colour

Q2 This stem and leaf diagram shows the exam scores of a group of Year 9 pupils.

  a) How many pupils got a score between 60 and 70? ....

  b) How many scored 80 or more? ....

  c) What was the most common test score? ....

  d) How many scored less than 50? ....

  e) How many pupils took the test? ....

```
3 | 2 3
4 | 6 8 8
5 | 1 2 2 3 6 6 9
6 | 1 5 5 5 8
7 | 2 3 4 5 8
8 | 0 1 1 5
9 | 0 2 3
```

Key: 5 | 2 means 52

This is nothing new... you've seen it all before. All you do is read off the information from the chart.

4.8 Questions on Bar Charts

<u>Remember</u> — the total of all the bars is the total frequency.

Q3   This bar chart shows the marks from a test done by some students:

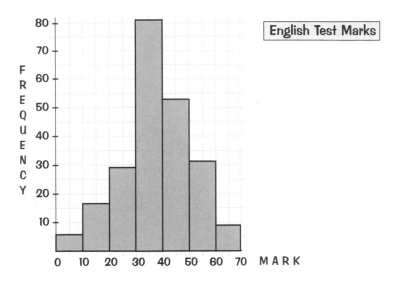

English Test Marks

a) How many students scored 20 marks or less?   ....

b) The pass mark for this test was 30. How many students passed the test?

....

c) How many students took the test?

....

The next bar chart shows the heights of a group of students.

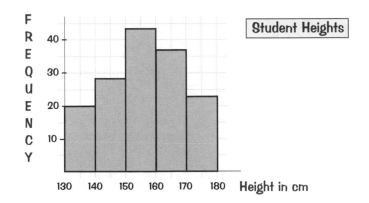

Student Heights

d) How many students are taller than 160 cm?   ....

e) How many students are between 140 and 160cm tall? ....

f) Write your answer to part **e)** as a percentage of all the students who were measured.

....

SECTION FOUR — STATISTICS AND GRAPHS

4.9 Questions on Line Graphs

Q1 Billy took his temperature and recorded it on this graph.

What was his temperature at:

a) 10am ....

b) 2pm ....

c) What was his highest temperature
....

d) When was this ....

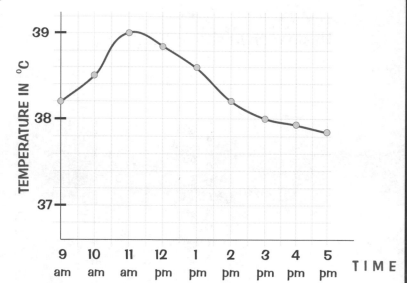

Q2 A baby was weighted every 5 days. The results are given below. Draw a graph to show how the baby's weight changed.

DAY N°	0	5	10	15	20	25	30
WEIGHT KG	5.3	5.2	5.9	6.4	6.6	6.7	6.8

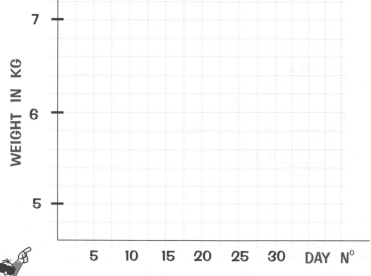

In your own words describe how the baby's weight changed:

....

....

....

....

Just draw in the points, then join them up with straight lines.

4.9   *Questions on Line Graphs*

I wonder why some of these graphs have dotted lines...

Q3 The number of pupils at a school over a period of years is shown in the graph below.

In which year was the number of pupils equal to:

a) 580 ....

b) 560 ....

How many pupils in:

c) 1994 ....

d) 1991 ....

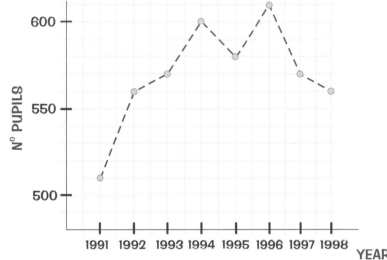

Q4 This table shows the number of cars in a city car park over a week. Draw a graph to show this information.

DAY	MON	TUE	WED	THU	FRI	SAT	SUN
N° CARS	32	36	18	33	37	40	6

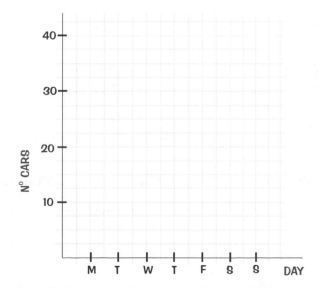

For the graphs on this page, why should the lines be dotted?

....

....

....

....

4.10 Questions on Scattergraphs

 Scatter graphs aren't supposed to make a nice line — they're always a bit messy. Just a load of points scattered round all over the shop.

Q1 These are the shoe sizes and heights for 12 pupils in Year 11.

Shoe size	5	6	4	6	7	7	8	3	5	9	10	10
Height (cm)	155	157	150	159	158	162	162	149	152	165	174	178

On the grid below draw a scattergraph to show this information.

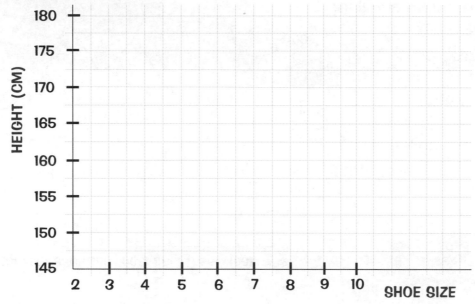

What does the scattergraph tell you about the relationship between shoe size and height for these pupils?

....

Q2 The scattergraphs below show the relationship between:

a) The temperature and the amount of ice cream sold.
b) The price of ice cream and the amount sold.
c) The age of people and the amount of ice cream sold.

Describe the correlation of each graph and say what each graph tells you.

a) ....

b) ....

c) ....

SECTION FOUR — STATISTICS AND GRAPHS

4.10 Questions on Scattergraphs

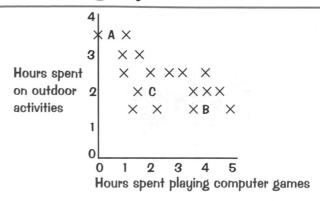

Q3 This scattergraph shows how much time a group of teenagers spend on outdoor activities and playing computer games.

Which of the points A, B or C represent each of these statements?

a)

The rugby practice was a long one so I didn't have much time to play on the computer.

Point ....

b) Point ....

I don't have a computer!

c) Point ....

I went to visit a friend. We played a bit of football then spent most of the evening playing his new computer game.

Q4 Alice wanted to buy a second hand car. She looked in the local paper and wrote down the ages and prices of 15 cars. On the grid below draw a scattergraph for Alice's information.

Age of car (years)	Price (£)
4	4995
2	7995
3	6595
1	7995
5	3495
8	4595
9	1995
1	7695
2	7795
6	3995
5	3995
1	9195
3	5995
4	4195
9	2195

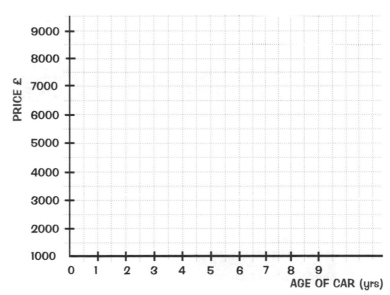

What does the scattergraph tell you about the relationship between the age of a car and its price? ....

The big word you're supposed to use in these questions is **CORRELATION** — and they're very keen on it, so make sure you know what it means.

4.11   *Questions on Pie Charts*

Q1   At a University there were 180 students in the Media Studies department.

Country	UK	Malaysia	Spain	Others
Number of students	90	35	10	45

To show this on a Pie chart you have to work out the angle of each sector. Complete the table showing your working. The UK is done for you.

COUNTRY	WORKING	ANGLE in degrees
UK	90 ÷ 180 × 360 =	180°
MALAYSIA		
SPAIN		
OTHERS		

Now complete the Pie chart using an angle measurer. The UK sector is done for you.

Q2   On TV, programmes of different types have the amount of air time as shown in the table.

Programme	Hours	Angle
News	5	
Sport	3	
Music	2	
Current Affairs	3	
Comedy	2	
Other	9	
Total	24	

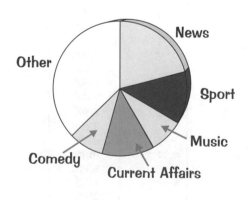

Complete the table by finding the size of the angle represented by each type of programme. Use an angle measurer or calculation method.

 The full circle (that's all 360° of it) represents the total of everything — so you shouldn't find any gaps in it, basically.

4.11   *Questions on Pie Charts*

Just follow the working in the questions — then **LEARN THE METHOD.**

Q3   Pupils at a school were asked about their activities at weekends. The results are shown in the table. Complete the table and then draw the pie chart using an angle measurer.

ACTIVITY	HOURS	WORKING	ANGLE
Homework	6	$6 \div 48 \times 360 =$	45
Sport	2		
TV	10		
Computer games	2		
Sleeping	18		
Listening to music	2		
Paid work	8		
Total	48		

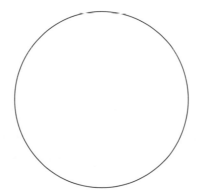

Q4   A family spends £540 each week on various items which are listed in the table and shown as sectors on the pie chart. Using an angle measurer or by calculation find the angle of each sector and enter it in the table

Item	£	Angle
Mortgage	150	
Heat/light	30	
Food	90	
Clothes	30	
Car	45	
General	195	

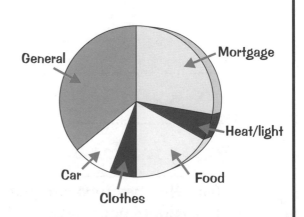

4.12 *Questions on Surveys*

Q1   Brian has been asked to conduct a survey into the different selections made by boys and girls of varying ages at the school midday meal break. Use the space below to design a tick box sheet which may help record his observations. (He will not ask any questions).

Q2   Sally has been asked to conduct a survey into the different television viewing habits of pupils at her school. List 4 questions which you believe may be helpful.

a) ....

b) ....

c) ....

d) ....

Yes / No questions are your best option — there's no grey area, so it's easier to evaluate your results... failing that, try ones that only ask for a number.

4.12   Questions on Surveys

If you want to know what people really think, you've got to ask the right questions — it's no use putting words in people's mouths.

Q3   John wanted to find out what was the most popular type of takeaway meal. He stood outside his local pizza shop and carried out a survey by asking the people going in and out their opinions. What is the most obvious fault with his survey?

....

....

Q4   Carol was asked to find out if most people have a calculator in their Maths lessons. She asked the people she liked in her Maths set if they had brought a calculator to school, 2 said yes, 3 said no. From this she claims most people do not have a calculator. Give 3 criticisms of Carol's survey.

....

....

Q5   David carried out a survey about the popularity of different subjects at school, among the questions he asked were:

a) Do you like games when it is raining?
b) Maths is much better than English isn't it?
c) Which subject do you spend more time on, Science or Drama?

Give a criticism of each question.

a) ....

b) ....

c) ....

4.13   *Questions on Probability*

Probability always has a value between 0 and 1 — if it's 0, the thing's **DEFINITELY NOT** going to happen... if it's 1, it **DEFINITELY IS** going to happen.

Q1   Write down whether these events are impossible, unlikely, even, likely or certain.

     a) You will go shopping on Saturday. ....

     b) You will live to be 150 years old. ....

     c) The next person who comes into the room is female. ....

     d) There will be a moon in the sky tonight. ....

Q2   Put an arrow on the probability line to show the following, also explain why you've put your arrow there:

     a)

       0   1

     The probability it will rain tomorrow.

     Explain: ....

     b)

       0   1

     The probability that Elvis Presley will release a new number one record.

     Explain: ....

Q3   Mike and Nick play a game of pool. The probability of Nick winning is 7/10.

     a) Put an arrow on the probability line below to show the probability of Nick winning. Label this arrow N.

     b) Now put an arrow on the probability line to show the probability of Mike winning the game. Label this arrow M.

     0   1

4.13   Questions on Probability

Q4   Write down, as a fraction, the probability of these events happening...

    a) Throwing a 5 with a dice. ....
    b) Drawing a red card from a pack of cards. ....
    c) Drawing a King from a pack of cards. ....
    d) Throwing a 0 with a dice. ....

Q5   Write down the probability of these events happening...

    a) Throwing an odd number with a dice. ....
    b) Drawing a black card from a pack. ....
    c) Drawing a Black King from a pack of cards. ....
    d) Throwing a prime number with a dice. ....

Q6   A bag contains ten balls. Five are red, three are yellow and two are green.
    What is the probability of picking out :

    a) A yellow ball. ....
    b) A red ball. ....
    c) A green ball. ....
    d) A red or a green ball. ....
    e) A blue ball. ....

Q7   A selection box contains 13 bars of chocolate. 6 are milk chocolate, 4 are dark
    chocolate and 3 are white chocolate. What is the probability that a bar picked
    at random contains...

    a) Dark chocolate. ....   **b)** Milk chocolate. ....
    c) No white chocolate. ....

Q8   Anne, Brett, Colin, Dan, Emma are trampolining. They can only go on in pairs.

    a) Write down all the different combinations of pairs that there are.
    (10 altogether).

    ....

    ....

    b) What is the probability that the pair does not contains a boy and a girl?

    ....

Remember — if they're asking the probability of something **NOT** happening,
it's the same as asking for the probability that the **OTHER** things **WILL** happen.

94

4.13 Questions on Probability

Q9 The outcome when a coin is tossed is head (H) or tail (T). Complete this table of outcomes when two coins are tossed together.

a) How many possible outcomes are there? ....
b) What is the probability of getting 2 heads? ....
c) What is the probability of getting a head followed by a tail? ....

		2ⁿᵈ COIN	
		H	T
1ˢᵗ COIN	H		
	T		

Q10 Two dice are rolled together. The scores on the dice are added. Complete this table of possible outcomes.

How many possible outcomes are there? ....

		SECOND DIE					
		1	2	3	4	5	6
FIRST DIE	1						
	2	3					
	3						
	4						
	5			8			
	6						

What is the probability of scoring:
a) 2 ....
b) 6 ....
c) 10 ....
d) More than 9 ....
e) Less than 4 ....
f) An even number ....
g) More than 12 ....

Q11 Two spinners are spun and the scores are multiplied together.

Fill in this table of possible outcomes.

What is the probability of scoring 12? ....

To win you have to score 15 or more. What is the probability of winning ?
....

		SPINNER 1		
		2	3	4
SPINNER 2	3			
	4			
	5			

  You're always better off using a table to put down the "possible outcomes" — you can't miss any out that way.

SECTION FOUR — STATISTICS AND GRAPHS

4.13   *Questions on Probability*

Well, OK, the probability is that you'd rather not be doing these at all...
still — this is the last page, so I'm sure you'll cope for a bit longer.

Q12 One day Sarah did a survey in her class on sock colour. She found out that pupils were wearing white socks, black socks or red socks. Jack said "If I pick someone at random from the class, then the probability that they are wearing red socks is 1/3." Explain why Jack might be wrong.

....

....

Q13 Imagine you have just made a 6-sided spinner in Design and Technology. How could you estimate the probability that it was a fair spinner?

....

....

....

Q14 How could you estimate the probability that it is going to snow on Christmas Eve this year?

....

....

....

Q15 "There is a 50% chance that it will rain tomorrow because it will either rain or it won't rain." Is this statement true or false? Explain your answer.

....

....

....

4.14 Questions on Coordinates

Q1 On the grid plot the following points. Label the points A,B... Join the points with straight lines as you plot them.

A(0,8) B(4,6) C(4.5,6.5) D(5,6) E(9,8) F(8,5.5) G(5,5) H(8,4) I(7.5,2) J(6,2) K(5,4) L(4.5,3)   M(4,4) N(3,2) O(1.5,2) P(1,4) Q(4,5) R(1,5.5) S(0,8).

You should see the outline of an insect. What is it?..............................

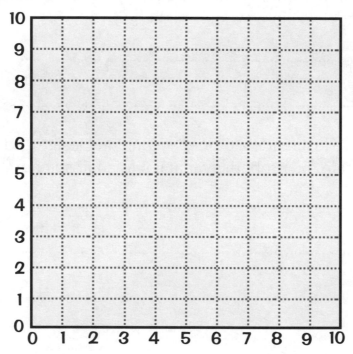

Q2 Write down the letter which is by each of the following points. The sentence it spells is the answer to question one.

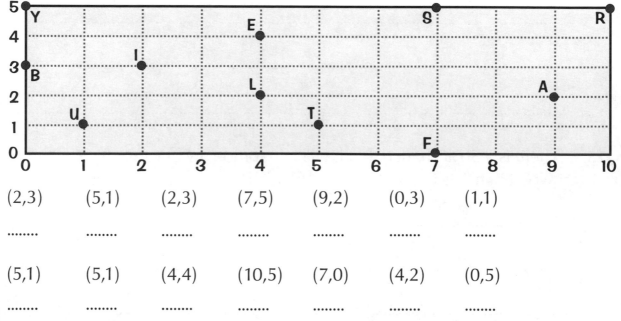

(2,3)   (5,1)   (2,3)   (7,5)   (9,2)   (0,3)   (1,1)

....

(5,1)   (5,1)   (4,4)   (10,5)   (7,0)   (4,2)   (0,5)

....

You've got to get your coordinates in the right order — or they're totally useless — you always go **IN THE HOUSE** then **UP THE STAIRS.**

4.14   *Questions on Coordinates*

Remember — 1)   **X comes before Y**

2)   **X goes a-cross** (get it) **the page.** (Ah, the old ones are the best...)

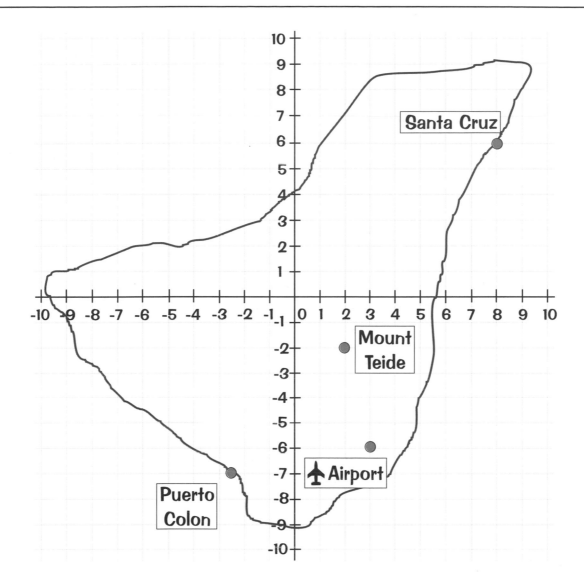

Q3   Here on the holiday island of Tenerife the sun always shines. Some important places are marked. What are their coordinates?

Airport (,)   Mount Teide (,)   Santa Cruz (,)   Puerto Colon (,)

Q4   Use the following coordinates to locate these holiday spots. Put the place name on the map.

Las Americas (-4 , -6)   El medano (4 , -4)   Icod (-6 , 2)   Laguna (3 , 7)
Taganana (9 , 9)

Q5   The cable car takes you to the top of Mount Teide. It starts at (3 , 1) and ends at (2 , -2). Draw the cable car route on the map.

SECTION FOUR — STATISTICS AND GRAPHS

4.15   *Questions on Midpoints of Lines*

Q1   Find the midpoint of the line AB, where A and B have coordinates:

a) A(2,3)   B(4,5)   ....

d) A(3,15)   B(13,3)   ....

b) A(1,8)   B(9,2)   ....

e) A(6,6)   B(0,0)   ....

c) A(0,11)   B(12,11)   ....

f) A(15,9)   B(3,3)   ....

ahh... nice'n'easy...

Q2   Find the midpoints of each of these lines:

a) Line PQ, where P has coordinates (1,5)
and Q has coordinates (5,6).   ....

b) Line AB, where A has coordinates (3,3)
and B has coordinates (4,0).   ....

c) Line RS, where R has coordinates (4,5)
and S has coordinates (0,0).   ....

d) Line PQ, where P has coordinates (1,3)
and Q has coordinates (3,1).   ....

e) Line GH, where G has coordinates (0,0)
and H has coordinates (−6,−7).   ....

Q3   Find the midpoint of each of the lines on this graph.

AB: ....

CD: ....

EF: ....

GH: ....

JK: ....

LM: ....

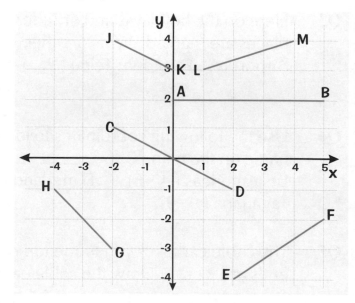

SECTION FOUR — STATISTICS AND GRAPHS

4.16   Questions on Graphs from Equations

With graphs from equations, you'll either get a straight line or a smooth curve
— so if you've got a point that looks wrong, it is. So do it again.

Q1   On the grid to the right draw the lines:

$y = 2$, $x = 5$, $y = 4$, $x = 0$, $y = -1$, $x = -2$.

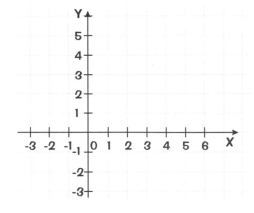

Q2 a) To draw the line $y = x + 5$:

i) Complete this pattern

x	(y) x + 5
-3	2
-2	
-1	
0	
1	6
2	
3	

ii) Write the coordinates

Coordinates
(-3,2)
(1,6)

....
....
....
....
....
....
....

iii) Plot the points on
the grid below and
join them up.

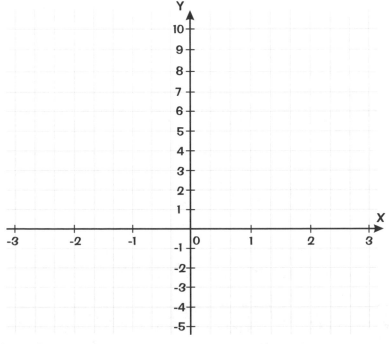

b) Repeat for the line $y = 2x + 3$.

5.1 Questions on Compass Directions

Q1

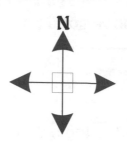

Start at the dot in the middle of the bottom line and follow the directions.
What shape have you drawn? ....

a) West 4 squares.
b) North 4 squares.
c) East 4 squares.
d) South 4 squares.
e) North East through 2 squares.

f) North 4 squares.
g) South West through 2 squares.
h) West 4 squares.
i) North East through 2 squares.
j) East 4 squares.

Q2

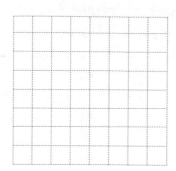
Joe's house

Shop

Church

Sue's house

Park

Jane's house

a) What direction does Jane go to get to Sue's house? ....

b) What direction is the church from Joe's house? ....

c) What is South East of Sue's house? ....

d) What is West of Sue's house? ....

e) Jane is at home. She is going to meet Sue in the park. They are going to the shop and then to Joe's house. Write down their directions.

....

You could use *"Never Eat Shredded Wheat"* but it's more fun to make one up — like *Naughty Elephants Squirt Water*, or *Nine Elves Storm Wales...* (hmm)

5.2   Questions on Three Figure Bearings

Bearings always have three digits — even the small ones... in other words, if you've measured 60°, you've got to write it as 060°.

This is a map of part of a coastline. The scale is one cm to one km.

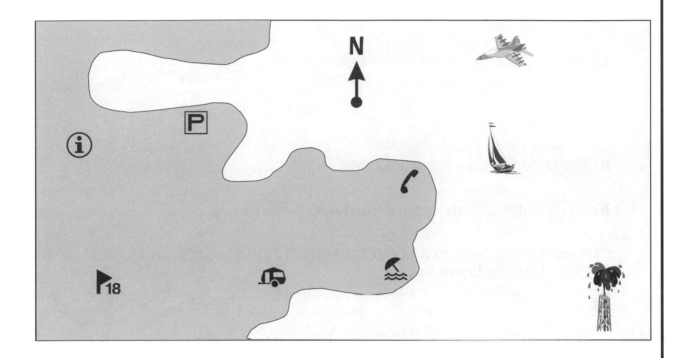

Q1   The water ski centre, 🎿, is on a bearing of 050° from the golf course, ⛳₁₈, and at a distance of 4.5 km. Put a ☆ where the ski centre is.

Q2   What is the bearing of the 🅿 from the 🏖? ....

Q3   What is the bearing of the 🚐 from the ⓘ? ....

Q4   How far and on what bearing is...

  a) The boat from the plane? ....

  b) The boat from the oil rig? ....

  c) The plane from the oil rig? ....

Q5   There is a lighthouse on the coast. It is at a bearing of 300° from the oil rig and a bearing of 270° from the boat. Mark its position with an ▲.

5.3 Questions on Scale Drawings

Q1 The scale on this map is 1cm : 4km.

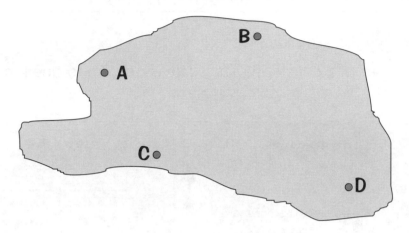

a) Measure the distance from A to B in cm. ....

b) What is the actual distance from A to B in km? ....

c) A helicopter flies on a direct route from A to B, B to C and C to D. What is the total distance flown in km?

....

Q2 Here is a plan of a garden drawn to a scale of 1 : 50.

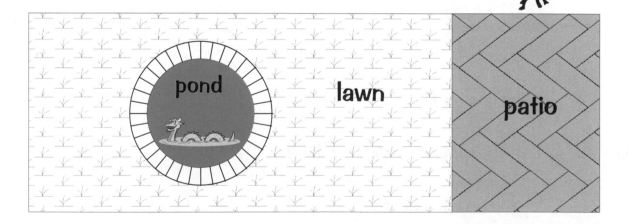

a) Measure the full length of the garden in mm: ....

b) What is the actual length of the garden in mm? ....

c) What is the actual length of the garden in metres? ....

 If the scale doesn't say what units it's in, it just means that both sides of the ratio are the same units — so _1 : 1000_ would mean _1cm : 1000cm_.

5.3 Questions on Scale Drawings

Watch out for those units... there's quite a mixture here — you'll
have to convert some of them before you can go anywhere.

Q3 A room measures 20m long and 15m wide. Work out the measurements for a
scale drawing of the room using a scale of 1cm = 2m.

Length = ...;   Width = ...

Q4 Katie drew a scale drawing of the top
of her desk. She used a scale of 1:10.
This is her drawing of the computer
keyboard. What are the actual
dimensions of it?

Length = ...;   Width = ...

Q5

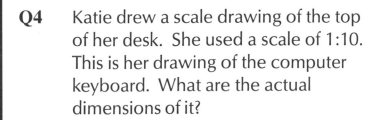

This is a scale drawing of Paul's new car.
Measure the length of the car. ....cm.
If the drawing uses a scale of 1:90,
work out the actual length of the car.

....

Q6 A rectangular field is 60m long and 40m wide. The farmer needs to make a
scale drawing of it. He uses a scale of 1:2000. Work out the measurements
for the scale drawing. (Hint — change the m to cm)

....

....

Q7 A room is 4.8m long and 3.6m wide. Make a scale drawing of it using a scale
of 1cm to 60cm. First work out the measurements for the scale drawing.

Length = ...
Width = ...

On your scale drawing
mark a window, whose
actual length is 2.4m, on
one long wall and mark a
door, actual width 90cm,
on one short wall.

Window = ...
Door = ...

5.4 *Questions on Travel Graphs*

Say SOD IT every time you come to one of these questions... that'll help you remember the order of Speed, Distance and Time in the formula triangle.

The graph below shows Nicola's car journey from her house to Alan's house (D), picking up Robbie (B) on the way.

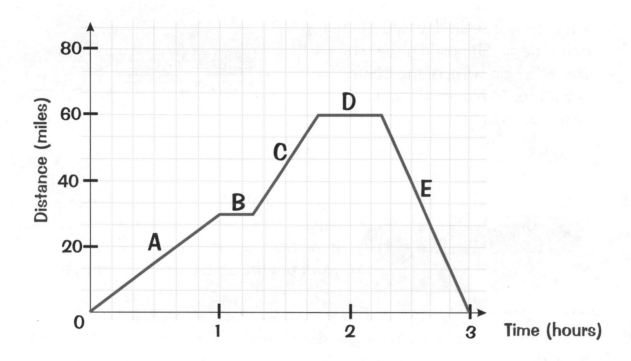

Q1   If Nicola started her trip at 10.00 am at what time does she return home?

....

Q2   How far is Robbie's house from Nicola's? ....

Q3   What is happening to the speed of the car along section B? ....

Q4   How long did they stop at Alan's for? ....

Q5   During which section was the speed greatest? ....

Q6   How long did the return journey take? ....

Q7   What is the speed of the car during section E? ....

SECTION FIVE — ANGLES AND OTHER BITS

5.5 Questions on Speed, Distance, Time

Q1 A car travels a total distance of 300 miles in 6 hours. What is the average speed for the journey?

Speed = Distance ÷ Time

Speed = ... ÷.............. = ... miles per hour.

Q2 It takes John 15 minutes to cycle to school, he lives 8 kilometres from the school. At what speed, in kilometres per hour, does he cycle?

15 minutes = 0.25 hours

Speed = ... = ... kph.

Q3 Hannah walks to the youth club, it takes her 20 minutes to walk 2 miles. Her friend Jess lives 6 miles away from the club, she gets there by car going, on average, 30 miles per hour.

a) At what speed does Hannah walk? ....

b) How long does it take Jess to get to the club?   **Time = Distance ÷ Speed.**

Time = ... ÷ .... = ... hours, which is .... × 60 = ... minutes.

Q4 The World Record for swimming 50m freestyle is 17.4 secs. What speed is this?

....

Q5 The speed of light is 18000000 metres per second. How many kilometres would light travel in 4 seconds?

....

Q6 a) If I leave my house to go to a party 18 miles away at 7.35pm, and I need to be at the party for 8.05pm. At what average speed would I have to drive?

....

b) Rob catches a train at 4.30pm to go to the same party, 150 miles away. Assuming that there are no delays and the train travels at an average speed of 75 mph. What time will Rob arrive at the party?

....

You need to **learn** the **different stages** of the travel graph, and what they look like — that's **TRAVELLING AWAY, TRAVELLING BACK** and **STATIONARY.**

5.6   *Questions on Determining Angles*

Estimating angles is easy once you know the 4 special angles — you can use them as reference points.

For each of the angles below write down its name, estimate its size (before you measure it!) and finally measure each angle with a protractor. The first one has been done for you.

Angle	Name	Estimated Size	Actual Size
a	acute	40°	43°
b			
c			
d			
e			
f			

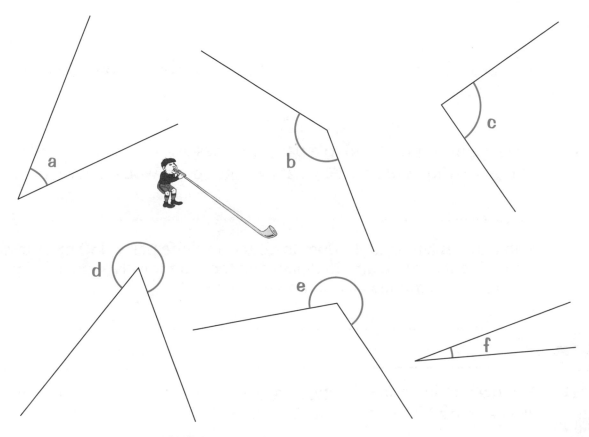

5.7 Questions on Drawing Angles

These instruments are used to measure angles.

An angle measurer.

A protractor.

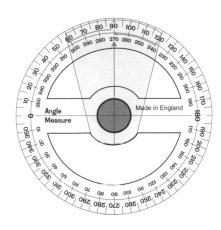 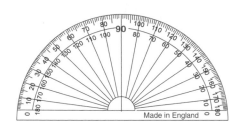

Q1 Use an angle measurer or protractor to help you to draw the following angles.

a) 20°

b) 65°

c) 90°

d) 136°

e) 225°

f) 340°

Q2 a) Draw an acute angle and measure it. **b)** Draw an obtuse angle and measure it.

Acute angle measures ....°

Obtuse angle measures ....°

c) Draw a reflex angle and measure it.

Reflex angle measures ....°

Don't forget protractors have two scales — one going one way and one the other... so make sure you measure from the one that starts with 0°, not 180°.

5.8   Questions on Angles

Hope you've learnt those angle rules for a straight line and round a point...

Q1   Work out the angles labelled:

a = ...

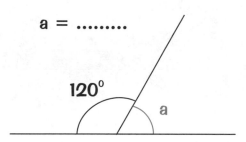

120°   a

b = ...

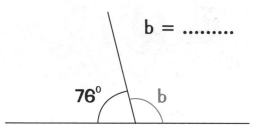

76°   b

c = ...

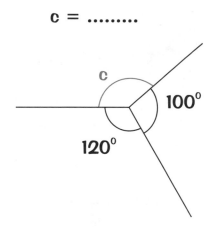

c   100°   120°

d = ...

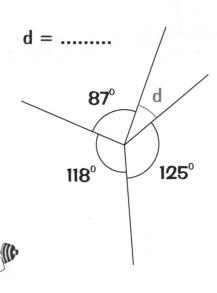

87°   d   118°   125°

e = ...

f = ...

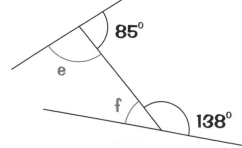

85°   e   f   138°

g = ...

h = ...

i = ...

g   41°   h   53°   i

SECTION FIVE — ANGLES AND OTHER BITS

5.8   *Questions on Angles*

Q2

> The three angles inside a triangle always add up to 180°

Work out the missing angle in each of these triangles. The angles are not drawn to scale so you cannot measure them.

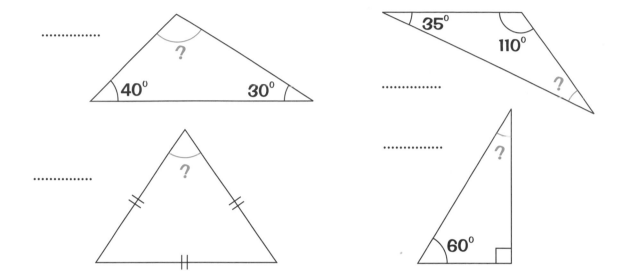

> The angles in a quadrilateral always add up to 360°

Q3   Work out the missing angles in these quadrilaterals.

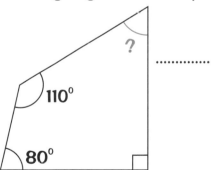

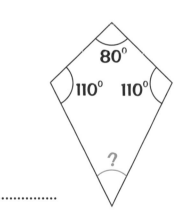

Q4   Work out the missing angles in these diagrams.

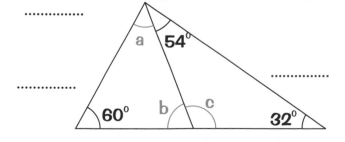

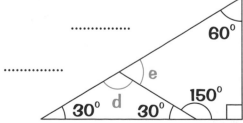

You'd better get learning these rules too — they're not that hard, and you'll be well and truly stumped without them.

5.8 Questions on Angles

More rules... once you know the **3 ANGLE RULES** for parallel lines, you can find all the angles out from just one — ah, such fun...

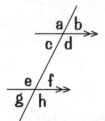

c = f and d = e — Alternate angles
a = e, c = g, b = f and d = h — Corresponding angles
$d + f = 180^0$, $c + e = 180^0$ — Supplementary angles

Q5 Find the sizes of the angles marked by letters in these diagrams. Write down what sort of angle each one is.

NOT DRAWN TO SCALE

a = ...

c = ...

b = ...

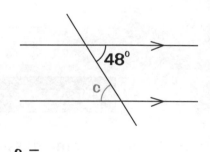

c = ...

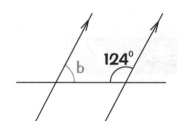

d = ...

e = ...

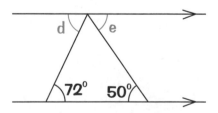

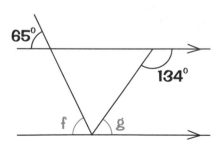

h = ...

f = ...

i = ...

g = ...

j = ...

5.8   *Questions on Angles*

Q6   Find the missing angles. The diagrams are not drawn to scale.

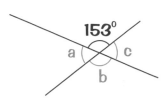 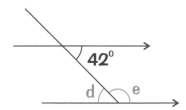 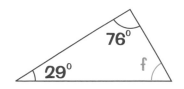

a =............°   b =............°   c =............°   d =............°   e =............°   f =............°

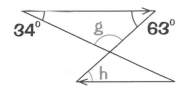 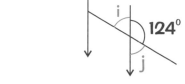 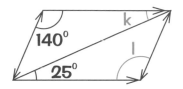

g=............°   h=............°   i=............°   j=............°   k=............°   l=............°

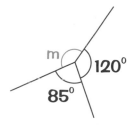  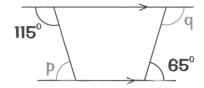

m=............°   n=............°   p=............°   q=............°

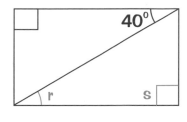 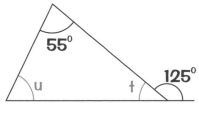 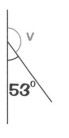

r=............°   s=............°   t=............°   u=............°   v=............°

Q7   How many different types of angle are in this picture? ....

What are the names of the angles you have found?

....

Look for the shapes you know the rules for (triangles, quadrilaterals, parallel lines, etc.) — use them and you can fill in the gaps to your heart's content.

SECTION FIVE — ANGLES AND OTHER BITS

5.9   *Questions on Polygons*

Q1   Here is a regular octagon:

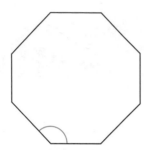

     a) What is the total of its eight interior angles?

         ....

     b) What is the size of the marked angle? ....

Here is an irregular hexagon:

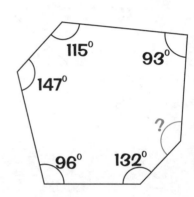

c) What is the total of its six interior angles? ....

d) What is the size of the missing angle? ....

Here is a regular pentagon:

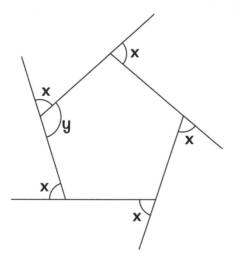

     e) The five angles marked x are its exterior angles. What do they add up to?

         ....

     f) Work out the value of x

         ....

     g) Use your answer from part f to work out the value of angle y

         ....

Here is a diagram showing part of a regular polygon:

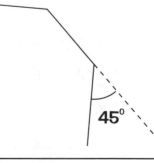

h) The angle shown is one of its exterior angles. From this, work out how many sides the polygon has.

     ....

A polygon can have <u>any</u> number of sides, but its <u>exterior</u> angles will always add up to 360° — ain't that just something...

5.9 Questions on Polygons

Even more rules — you've just gotta get down and learn them... and here's another: INTERIOR ANGLE + EXTERIOR ANGLE of <u>any</u> POLYGON = 180°.

Q2 The diagram shows a regular hexagon:

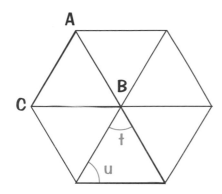

a) Work out the size of angles t and u. ....

b) What type of triangle is ABC? ....

This hexagon has equal sides but not equal angles.

c) The two angles marked v are equal.
Calculate these angles.

....

A decagon has ten sides.

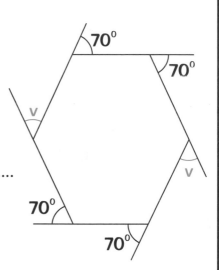

d) What do its 10 exterior angles add up to? ....

If the decagon is regular...

e) What size is each exterior angle? ....

f) what size is each interior angle? ....

g) If the interior angles of a polygon add up to 3240°,
how many sides has it got?

....

Q3 A maths teacher walks 10 paces forward. Then she turns 30° clockwise and walks 10 paces forward. She repeats this process until she ends up back where she started. What type of polygon has she traced out in her walk?

....

SECTION FIVE — ANGLES AND OTHER BITS

5.10 Questions on Congruence and Similarity

I reckon these are pretty easy — so while you're racing through them, you can be thinking: "3 shapes are CONGRUENT (exactly the same) and 1 isn't."

In each of the following sets of shapes underline the one which is not congruent.

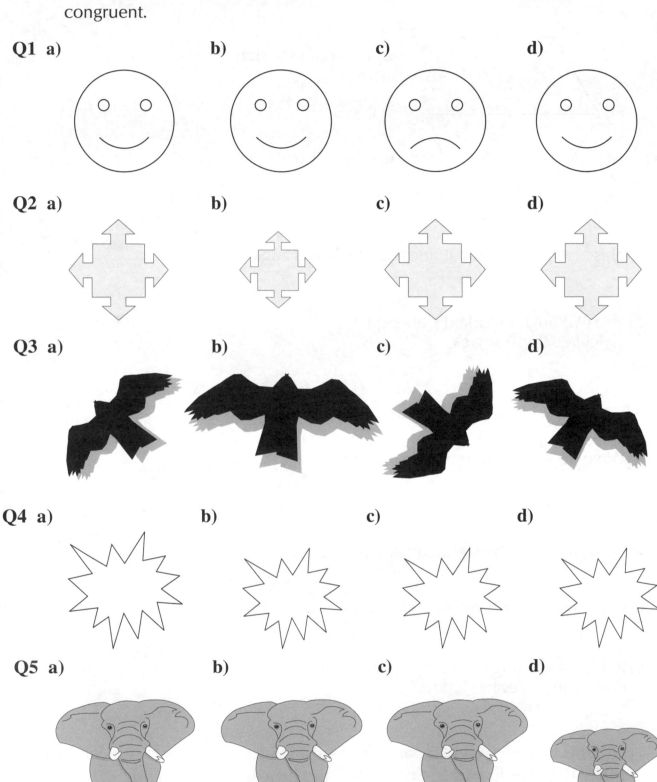

Q1 a)   b)   c)   d)

Q2 a)   b)   c)   d)

Q3 a)   b)   c)   d)

Q4 a)   b)   c)   d)

Q5 a)   b)   c)   d)

5.10 Questions on Congruence and Similarity

Q6   Here are 3 rectangles, which 2 are similar?

a)
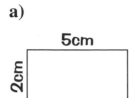
5cm
2cm

b)
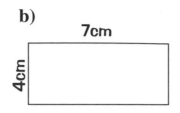
7cm
4cm

c)
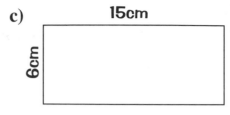
15cm
6cm

....

Q7   Here are 3 triangles, which 2 are similar?

a)
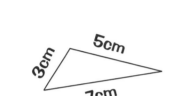
5cm
3cm
7cm

b)
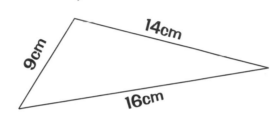
14cm
9cm
16cm

c)
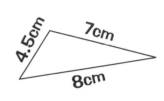
7cm
4.5cm
8cm

....

Q8   Which of the following rectangles are similar to this one?
(Yes/No)

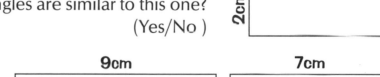
6cm
2cm

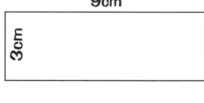
5cm
1cm

3cm
1cm

9cm
3cm

7cm
3cm

....

Q9   These 2 shapes are similar. Find the labelled lengths.

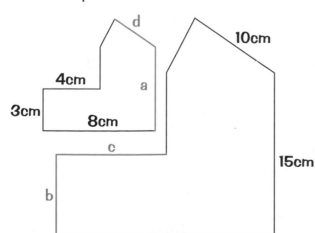
d
10cm
4cm
a
3cm
8cm
c
15cm
b
20cm

a) ....

b) ....

c) ....

d) ....

Shapes are SIMILAR if they are the SAME SHAPE but DIFFERENT SIZE.
Let's face it, it's a much better word than congruent...

SECTION FIVE — ANGLES AND OTHER BITS

5.11 Questions on Enlargement

The scale factor is a fancy way of saying **HOW MUCH BIGGER** the enlargement is than the original.

Q1 What is the scale factor of each of these enlargements?

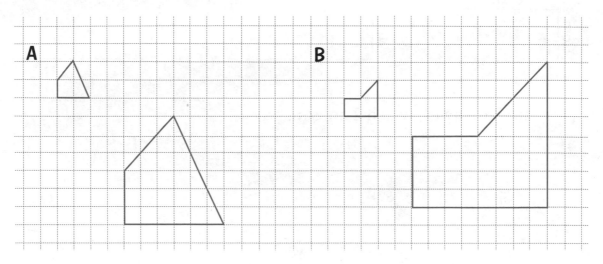

    a) Scale factor is ....   **b)** Scale factor is ....

Q2 Enlarge these figures with scale factor 2.

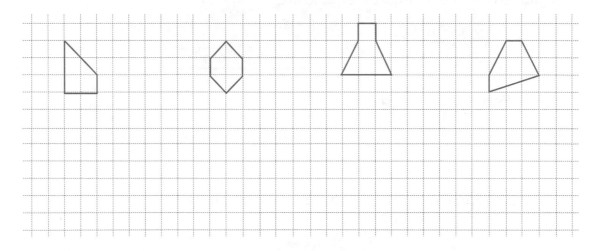

Q3 Enlarge this triangle by scale factor 3 with O as the centre of enlargement.

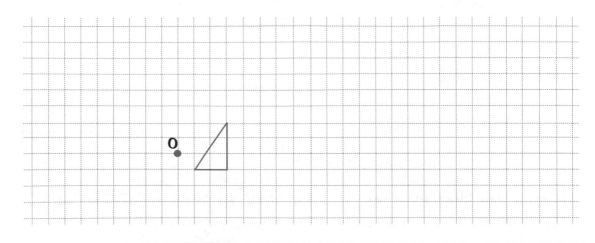

5.11 — Questions on Enlargement

Q4 Enlarge each of these diagrams by scale factor 2, using the point (0,0) as the centre of enlargement.

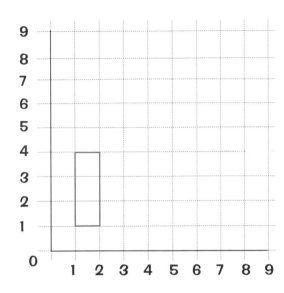 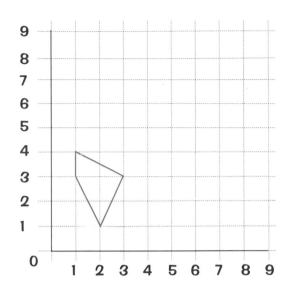

Q5 Enlarge this shape using scale factor 3 and centre of enlargement E.

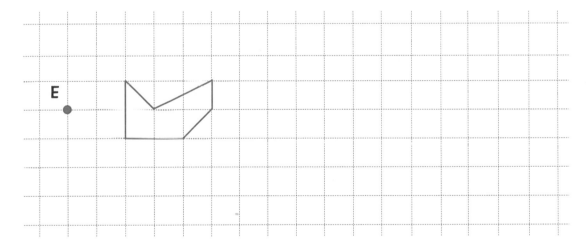

Q6 Enlarge this triangle using scale factor 4 and centre of enlargement C.

Draw a line through the _centre of enlargement_ to each point, then make that line longer by whatever the _scale factor_ is to get the new point.

5.12   Questions on Rotation

Q1   The centre of rotation for each of these diagrams is **X**. Rotate (turn) each shape as asked then draw the new position of the shape onto each of the diagrams below.

a) 180° clockwise (or ½ turn).

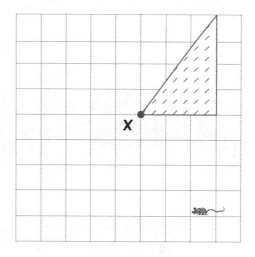

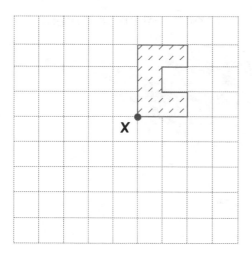

b) 90° clockwise (or ¼ turn clockwise).

c) 270° anticlockwise (or ¾ turn anticlockwise).

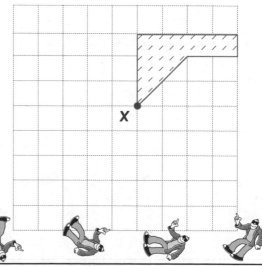

 A _½ turn clockwise_ is the same as a _½ turn anti-clockwise_ — and a _¼ turn clockwise_ is the same as a _¾ turn anti-clockwise_. Great fun, innit...

SECTION FIVE — ANGLES AND OTHER BITS

5.12   *Questions on Rotation*

You can check what it's supposed to look like by spinning your book round with your pen on the centre of rotation.

Q2   The centre of rotation is the origin O.

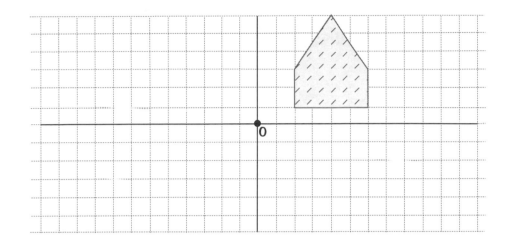

   a) Rotate the shaded shape 90° clockwise. Label the new image **A**.

   b) Rotate the shaded shape 180° clockwise. Label the new image **B**.

   c) Rotate the shaded shape 270° clockwise. Label the new image **C**.

   d) Through how many degrees clockwise would you turn image **C** to return to return into the position of the shaded shape?

                                           ....

Q3   This is a scalene triangle PQR. The centre of rotation is the origin O.

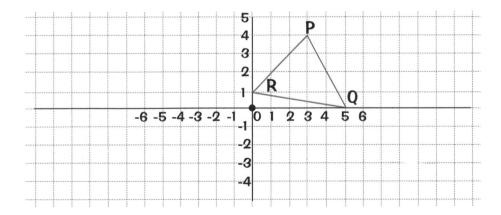

   a) Write down the coordinates of...

     P ....   Q ....   R ....

   b) Rotate the triangle 90° anticlockwise about the origin 0. Label the new triangle P′ Q′ R′.

   c) Write down the coordinates of ...

     P′ ....   Q′ ....   R′ ....

5.13   *Questions on Reflection*

Q1   Reflect each shape in the line x = 4.

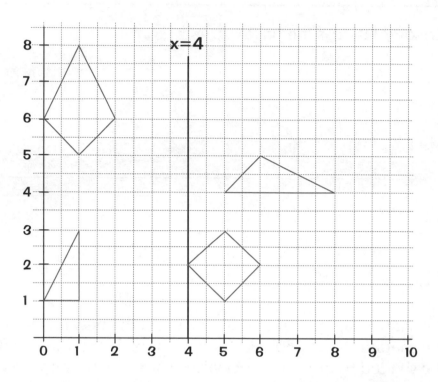

Q2   Reflect each shape in the line, y = 4.

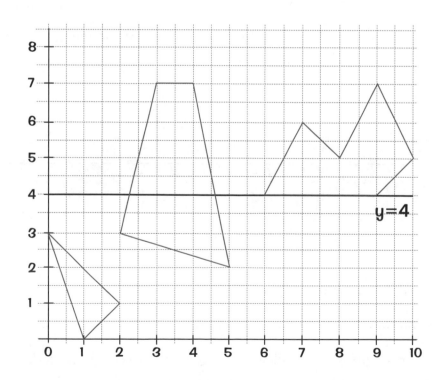

Nothing fancy here, is there — reflection's just mirror drawing really. And we've all done that before... in fact we did it in Section 2, if you remember.

SECTION FIVE — ANGLES AND OTHER BITS

5.13 *Questions on Reflection*

It's just the same as the ones you did in the symmetry bit — so remember that trick about using tracing paper.

Q3 Reflect the shapes in the line y = x.

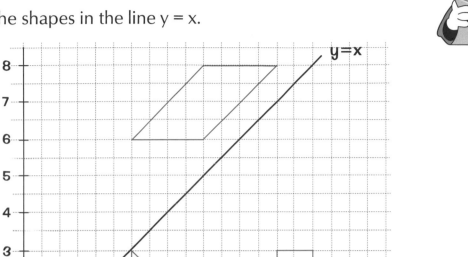

Q4 Reflect ① in the line y = 5, label this ②. Reflect ② in the line x = 9, label this ③. Reflect ③ in the line y = x, label this ④. Reflect ④ in the line x = 4, label this ⑤. Reflect ⑤ in the line y = x, label this ⑥.

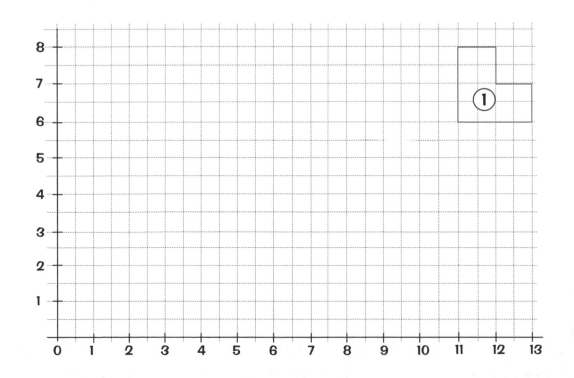

5.14   *Questions on Translation*

The movement along a straight line in a particular direction.
The shape slides across from one position to another.

Q1   The arrow is to be translated 10 squares right then 2 squares up. Draw the image.

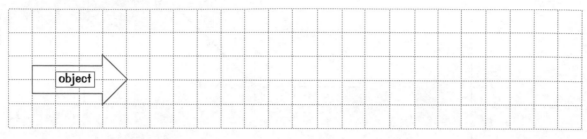

Q2   Translate the following shapes and draw the images.
Label the images A′ B′ C′

A   4 left , 3 down   B   5 right, 5 up   C 4 right, 4 down

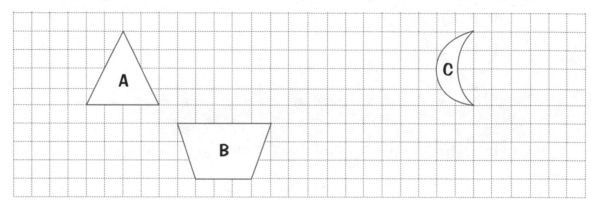

Q3   Translate this shape, drawing the image each time:
(3 right, 4 up) then (9 right, 2 up) then (3 right, 4 down) then (8 left, 4 down)
Label the images A′, A″ , A‴, A⁗.

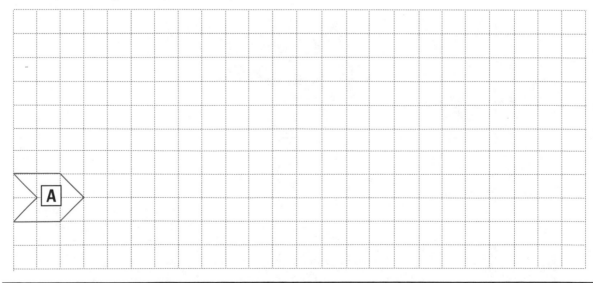

This has got to be the easiest of the lot — you don't have to change the shape at all — all it does is move along a bit, or up a bit... or both.

SECTION FIVE — ANGLES AND OTHER BITS

5.14 Questions on Translation

When you've got to say what the translation is, you only need to follow what one point is doing — the rest of the shape will do exactly the same.

Q4 Look at the following diagram and describe the translations.

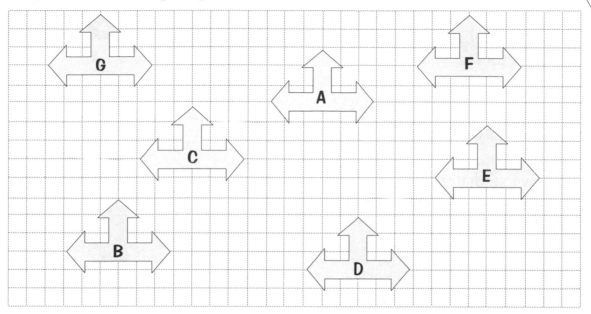

G to E ....   A to F ....

E to A ....   C to D ....

D to C ....   B to G ....

F to B ....

Make a comment on the translations (D to C) and (C to D) ....

Q5

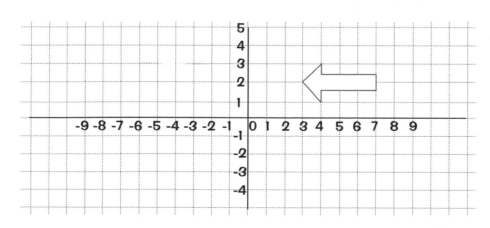

The arrow is translated (10 left, 4 down). Draw the arrow in its new position. The coordinates of the tip of the arrow are (3 , 2). The new coordinates of the tip are (,   ).

5.15 Questions on Constructions

Constructions should always be done as accurately as possible using: sharp pencil, ruler, compasses, protractor (set-square).

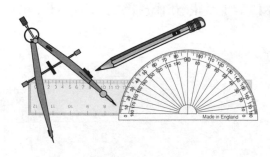

Q1 a) Draw a circle with radius 4 cm.

Draw in a diameter of the circle. Label one end of the diameter X and the other end Y.

Mark a point somewhere on the circumference — not too close to X or Y. Label your point T. Join X to T and T to Y.

Measure angle XTY.   XTY = ...°

b) Make an accurate drawing below of the triangle on the right. Measure side AB on your triangle giving your answer in millimetres.

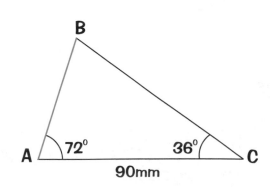

AB = ... mm

5.15 *Questions on Constructions*

Q2 a) Draw below an accurate full-size version of the triangle on the right. Measure side PQ on your triangle giving your answer in millimetres.

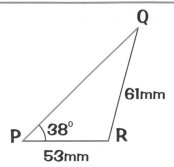

P •

PQ = ... mm

b) Draw below an accurate full-size version of the square on the right.

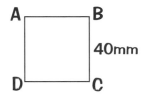

On your diagram mark the midpoint of BC and label it M. Now draw a circle radius 20mm with M as its centre. Next join A to M and where it crosses the circle label X. Now join D to M and where it crosses the circle label Y. Join X to Y. Now measure line XY.   XY= .... mm
Line XY is called a .... of the circle.
Triangle MXY is called an .... triangle.

A
•

Yeah, these do get a bit tricky, I admit, but just follow the instructions and draw neatly with a (sharp) pencil and you won't go far wrong.

6.1   *Questions on Powers*

You've seen squares and cubes before (come on, it wasn't that long ago).
This is more or less the same — except you multiply a few more times.

Q1   Work out the square of these numbers.

a) 3 ....   **b)** 12 ....   **c)** 15 ....

Q2   What is the cube of these numbers?

a) 4 ....   **b)** 7 ....   **c)** 20 ....

Q3   Find the value of these:

a) $4^2 =$ ....   **b)** $16^2 =$ ....   **c)** $3^3 =$ ....

d) $6^3 =$ ....   **e)** $2^5 =$ ....   **f)** $5^4 =$ ....

g) $1^7 =$ ....   **h)** $2^7 =$ ....   **i)** $3^5 =$ ....

Q4   Write these in index notation.

a) $5 \times 5 \times 5 \times 5 =$ ....   **b)** $3 \times 3 \times 3 =$ ....

c) $7 \times 7 =$ ....   **d)** $11 \times 11 \times 11 \times 11 \times 11 \times 11 =$ ....

e) six squared = ...   **f)** eight cubed = ...

Q5   Work out:

a) $7^2 - 6^2 =$ ....   **b)** $8^2 - 7^2 =$ ....

c) $9^2 - 8^2 =$ ....   **d)** $10^2 - 9^2 =$ ....

What do you notice? ....

Use your last answer to work out $30^2 - 29^2$ without using a calculator. ....

Q6   Complete this number pattern:

1^3 = ...
$1^3 + 2^3$ = ...
$1^3 + 2^3 + 3^3$ = ...
$1^3 + 2^3 + 3^3 + 4^3 =$ ....

What do you notice about your answers? ....
....

Q7   Your calculator may have a button for working out powers.
It looks like this [Xʸ] .
e.g. 5^3 On your calculator key in [5] [Xʸ] [3] [=] .
You should get the answer 125.

Try these

a) $7^9 =$ ....   **b)** $2^{12} =$ ....

c) $23^5 =$ ....   **d)** $13^7 =$ ....

6.2   *Questions on Square Roots*

Q1   What are the answers to the following...

     a) $\sqrt{16}$ = ...
       b) $\sqrt{49}$ = ...
       c) $\sqrt{81}$ = ...

     d) $\sqrt{100}$ = ...
       e) $\sqrt{1}$ = ...
       f) $\sqrt{144}$ = ...

Q2   The square root of 10 must be between the square root of 9, which is 3, and the square root of 16, which is 4. Therefore $\sqrt{10}$ is between 3 and 4. Complete the following...

     a) $\sqrt{1}$ = ..., $\sqrt{4}$ = ..., so $\sqrt{2}$ is between .... and ....

     b) $\sqrt{16}$ = ..., $\sqrt{25}$ = ..., so $\sqrt{20}$ is between .... and ....

     c) $\sqrt{50}$ is between .... and ....

     d) $\sqrt{115}$ is between .... and ....

     e) $\sqrt{150}$ is between .... and ....

Q3   Use your calculator to answer the following. Give your answers correct to two decimal places.

     a) $\sqrt{41}$ = ...
       b) $\sqrt{75}$ = ...
       c) $\sqrt{106}$ = ...

     d) $\sqrt{137}$ = ...
       e) $\sqrt{181}$ = ...
       f) $\sqrt{200}$ = ...

     g) $\sqrt{225}$ = ...
       h) $\sqrt{250}$ = ...
       i) $\sqrt{1000}$ = ...

Q4   If a square has an area of 64cm², what is the length of its sides?

           Area = side²
           side² = 64
           so, side = ... cm

A square root just means WHAT NUMBER TIMES ITSELF GIVES... It's just the reverse of squaring, in fact.

6.3 *Questions on Number Patterns*

Q1 Draw the next two pictures in each pattern. How many match sticks are used in each picture?

a)

....

b)

....

c)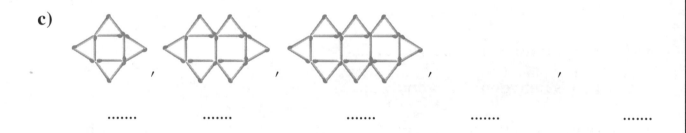

....

d)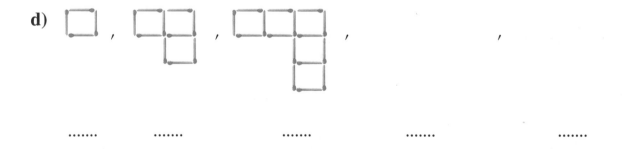

....

e)

....

Look for patterns in the numbers as well as pictures.

6.3   *Questions on Number Patterns*

Look for the pattern and then fill in the next three lines. Some of the answers are too big to fit on a calculator display so you must spot the pattern.

a)

$$7 \times 6 = 42$$
$$67 \times 66 = 4422$$
$$667 \times 666 = 444222$$
$$6667 \times 6666 = \text{..........................}$$
$$66667 \times 66666 = \text{..........................}$$
$$666667 \times 666666 = \text{..........................}$$

b)

$$99 \times 11 = 1089$$
$$99 \times 22 = 2178$$
$$99 \times 33 = 3267$$
$$99 \times 44 = \text{.............}$$
$$99 \times 55 = \text{.............}$$
$$99 \times 66 = \text{.............}$$

c)

$$1 \times 9 + 2 = 11$$
$$12 \times 9 + 3 = 111$$
$$123 \times 9 + 4 = 1111$$
$$1234 \times 9 + 5 = \text{.................}$$
$$12345 \times 9 + 6 = \text{.................}$$
$$123456 \times 9 + 7 = \text{.................}$$

d)

$$1 \times 9 = 9$$
$$21 \times 9 = 189$$
$$321 \times 9 = 2889$$
$$4321 \times 9 = \text{.................}$$
$$54321 \times 9 = \text{.................}$$
$$654321 \times 9 = \text{.................}$$

e)

$$101 \times 22 = 2222$$
$$101 \times 222 = 22422$$
$$101 \times 2222 = 224422$$
$$101 \times 22222 = \text{...................}$$
$$101 \times 222222 = \text{...................}$$
$$101 \times 2222222 = \text{...................}$$

f)

$$9 \times 9 = 81$$
$$99 \times 99 = 9801$$
$$999 \times 999 = 998001$$
$$9999 \times 9999 = \text{....................}$$
$$99999 \times 99999 = \text{....................}$$
$$999999 \times 999999 = \text{....................}$$

g)

$$9 \times 5 = 45$$
$$99 \times 55 = 5445$$
$$999 \times 555 = 554445$$
$$9999 \times 5555 = 55544445$$
$$99999 \times 55555 = \text{.....................}$$
$$999999 \times 555555 = \text{.....................}$$
$$9999999 \times 5555555 = \text{.....................}$$

h)

$$1 \times 81 = 81$$
$$21 \times 81 = 1701$$
$$321 \times 81 = 26001$$
$$4321 \times 81 = 350001$$
$$54321 \times 81 = \text{.....................}$$
$$654321 \times 81 = \text{.....................}$$
$$7654321 \times 81 = \text{.....................}$$

The patterns aren't hard — they just look bad because of the huge numbers.

6.3   *Questions on Number Patterns*

Q1   In each of the questions below, write down the next three numbers in the sequence and write the rule that you used...

    a) 1, 3, 5, 7, ..., ..., ... Rule ....

    b) 2, 4, 8, 16, ..., ..., ... Rule ....

    c) 3, 30, 300, 3000, ..., ..., ... Rule ....

    d) 3, 7, 11, 15, ..., ..., ... Rule ....

    e) 6, 15, 24, 33, ..., ..., ... Rule ....

    f) 19, 14, 9, 4, –1, ..., ..., ... Rule ....

Q2   These are well known number sequences. Name them and write down the next two terms...

    a) 1, 4, 9, 16, 25, 36, ..., ...   Name ....

    b) 1, 3, 6, 10, 15, 21, ..., ...   Name ....

    c) 1, 1, 2, 3, 5, 8, 13, ..., ...   Name ....

Q3   The letter n describes the position of a term in the sequence. For example, if $n = 1$, that's the 1st term...if $n = 10$ that's the 10th term and so on. In the following use the rule given to generate (or make) first 5 terms.

    a) $3n + 1$ so if $n = 1$ the 1st term is <u>**(3 x 1) + 1 = 4**</u>

      $n = 2$ the 2nd term is .... = ...

      $n = 3$ .... = ...

      $n = 4$ .... = ...

      $n = 5$ .... = ...

    b) $5n - 2$, when $n = 1, 2, 3, 4$ and 5
      produces the sequence ...., ..., ..., ..., ...

    c) n^2, when $n = 1, 2, 3, 4,$ and 5
      produces the sequence ...., ..., ..., ..., ...

    d) $n^2 - 3$, when $n = 1, 2, 3, 4,$ and 5
      produces the sequence ...., ..., ..., ..., ...

    e) $(n + 2) \div 2$, when $n = 1, 2, 3, 4,$ and 5
      produces the sequence ...., ..., ..., ..., ...

Once you've worked out what you think the next numbers should be, go back and write down exactly what you did — that will be the rule you're after.

6.3   *Questions on Number Patterns*

To find the "nth term" there's a really easy formula to help you...
so **LEARN IT** — you'll be up the creek in the Exam without it.

Q4   Find the next two terms in this sequence:

$$3 \quad 7 \quad 11 \quad 15 \quad \ldots\ldots \quad \ldots\ldots$$
$$\quad\quad +4 \quad +4 \quad +4$$

The difference between the terms is always ....

To find the 'n'th number in the sequence you can use the simple rule 4n – 1

4n   is n × 4 the difference between the terms.

–1   is found by subtracting the difference between the terms from the first
term 3 – 4.

What will the 10th number in the sequence be? ....

Q5   Find the next to terms in this sequence 8   13   18   23   ....

What is the difference between the terms? ....

Subtract the difference between the terms from the first term   ....

Write down the rule for finding the 'n'th term ....

What is the 20th term in the sequence? ....

Q6   In each of the following patterns find:
a) The next two terms of the sequence.
b) The difference between the terms.
c) Subtract the difference between the terms from the first term.
d) The rule for the 'n'th term.
e) The 50th term of the sequence.

6   10   14   18   .... ; **a)**....... **b)**....... **c)**......... **d)**............ **e)**.........

5   8   11   14   .... ; **a)**....... **b)**....... **c)**........ **d)**............ **e)**.........

3   11   19   27   .... ; **a)**....... **b)**....... **c)**........ **d)**............ **e)**.........

6   9   12   15   .... ; **a)**....... **b)**....... **c)**........ **d)**............ **e)**.........

Q7   How many squares are in the next diagram in the pattern ....

What is the rule for the 'n'th term of the sequence? ....

6.4 *Questions on Negative Numbers*

Q1 Write these numbers in the correct position on the number line below:

a) –4   3   2   –3   0   –5   1

b) Which temperature is lower (colder), 8°C or –4°C ? ....

c) Which temperature is 1° warmer than –24°C ? ....

Put the correct symbol, < or >, between the following pairs of numbers:

d) 4   –8   **e)** –6   –2   **f)** –8   –7

g) –3   –6   **h)** –1   1   **i)** –3.6 –3.7

j) Rearrange the following numbers in order of size, largest first:

   –2   2   0.5   –1.5   –8 = ...

k) If the temperature is 6°C but it then gets colder and falls by 11°, what is the new temperature?

....

l) One day in winter the temperature at 0600 was –9°C. By midday, it had risen to –1°C. How many degrees did the temperature rise?

....

Always draw a number line to count along, so you can see what you're doing.

6.4   *Questions on Negative Numbers*

When you've got really big numbers, just mark off the tens (or even just the hundreds) — you'll be there all day otherwise.

Q2   Work out:

a) 2 – 7 = ...

b) 4 – 18 = ...

c) 1 – 20 = ...

d) 12 – 14 = ...

e) 72 – 77 = ...

f) 3 – 100 = ...

Q3   Work out:

a) -6 + 1 = ...

b) -10 + 2 = ...

c) -8 + 8 = ...

d) -70 + 3 = ...

e) -100 + 13 = ...

f) -1000 + 1 = ...

Q4   Work out:

a) -3 – 2 = ...

b) -3 – 6 = ...

c) -13 – 3 = ...

d) -50 – 4 = ...

e) -2 – 19 = ...

f) -7 – 96 = ...

Q5   Work out:

a) -2 + 5 = ...

b) -6 + 10 = ...

c) -2 + 50 = ...

d) -12 + 100 = ...

e) -47 + 56 = ...

f) -49 + 98 = ...

6.4 Questions on Negative Numbers

Q6 Work out:

a) 3 + 2 = ...

b) 3 + ⁻2 = ...

c) 7 + 10 = ...

d) 7 + ⁻10 = ...

e) 4 + ⁻19 = ...

f) 3 + ⁻100 = ...

Q7 Work out:

a) 6 – 2 = ...

b) 6 – ⁻2 = ...

c) 4 – 9 = ...

d) 4 – ⁻9 = ...

e) 17 – ⁻6 = ...

f) 8 – ⁻43 = ...

Q8 Work out:

a) -4 + 6 = ...

b) -4 – ⁻6 = ...

c) -6 – ⁻20 = ...

d) -18 – ⁻9 = ...

e) -30 – ⁻6 = ...

f) -25 – ⁻25 = ...

Q9 Work out:

a) -8 + 2 = ...

b) -8 + ⁻2 = ...

c) -10 + ⁻5 = ...

d) -31 + ⁻3 = ...

e) -4 + ⁻27 = ...

f) -47 + ⁻29 = ...

If you minus a negative number, it's the same as adding a positive one.
Isn't it great...

6.5 Questions on the Language of Algebra

It's no big mystery — algebra is just like normal sums, but with the odd letter or two stuck in for good measure.

Q1 Write the algebraic expression for these:

a) Three more than x ....

b) Seven less than y ....

c) Four multiplied by x ....

d) y multiplied by y ....

e) Ten divided by b ....

f) A number add five ....

g) A number multiplied by two ....

h) Two different numbers added together ....

Q2 Steven is 16 years old. How old will he be in:

a) 5 years ....   **b)** 10 years ....   **c)** x years ....?

Q3 Tickets for a football match cost £25 each. What is the cost for:

a) 2 tickets ....
b) 6 tickets ....
c) y tickets ....

CGP Wanderers Football Club
Vs United Rovers FC
Comfy Seat
East stand lower bit
Row 20
Seat 104
£25.00

Q4 There are n books in a pile. Write an expression for the number of books in a pile that has:

a) 3 more books ....

b) 4 fewer books ....

c) Twice as many books ....

Q5 a) I have 6 CDs and I buy 5 more. How many CDs have I now? ....

b) I have 6 CDs and I buy w more. How many CDs have I now? ....

c) I have x CDs and I buy w more. How many CDs have I now? ....

Q6 a) This square has sides of length 3cm.
What is its perimeter? ....
What is its area? ....

3cm
3cm

b) This square has sides of length d cm.
What is its perimeter? ....
What is its area? ....

dcm
dcm

6.6   *Questions on Basic Algebra*

Q1   Complete the following...

a) ● + ● + ● + ● + ● = 5●

b) ● + ● + ● + ● = ...●

c) ✤ + ✤ + ✤ = ...✤

d) ☆ + ☆ + ☆ + ● + ● = ...☆ + ...●

e) ✤ + ✤ + ● + ● + ● + ☆ = ...✤ + ...● + ...☆

f) ☆ + ☆ + ☆ + ☆ + ☆ + ☆ + ✤ + ✤ + ✤ =..........☆ + ...✤

g) ✤ + ✤ + ✤ + ✤ – ✤ – ✤ = ...✤

h) ☆ + ☆ + ☆ + ☆ + ☆ + ● + ● – ☆ – ☆ – ● = ...☆ + ...●

Q2   Collect the like terms together.

a) $2x + 3x =$ ....

b) $5x - 4x =$ ....

c) $6x + 2y - 3x + y =$ ....

d) $10x + 3y + 2x - 3y =$ ....

e) $5x + 3y - 2z - 6y =$ ....

f) $-4z + 6x - 2y + 2z - 3y =$ ....

g) $15x - 4y + 3z - z - 11x + 5y - y - 4x + z =$ ....

Q3   Remember that $x \times x = x^2$. Collect like terms in the following...

a) $y \times y \times y =$ ....

b) $y \times x =$ ....

c) $x \times 2x =$ ....

d) $y \times y + x \times x \times x =$ ....

e) $p \times p + 2q \times q \times q =$ ....

f) $r \times r \times r + q^2 \times q \times 3p^2 =$ ....

Remember that when a letter's stuck to a number there's a hidden times sign, in other words **2y = 2 × y** so **2y + 3y = y + y + y + y + y = 5 × y = 5y**

6.6   *Questions on Basic Algebra*

The thing OUTSIDE the brackets multiplies EVERYTHING INSIDE the brackets. Forget that and you've had it.

Simplify by multiplying out the brackets. The first one is done for you.

Q1 a) 3(2a + 4)

3(2a + 4) means three 'lots' of (2a + 4) , so

3(2a) + 3(+ 4)

6a + 12

b) 4(3b + 6)

....

c) 2(2c – 3)

....

d) 5(7d + 3)

....

e) 3(e – 5)

....

Q2 a) -2(4a – 3)

-2(4a – 3) means two 'lots' of (4a – 3), so

-2(4a) – 2(-3)

-8a + 6

b) -6(4b – 2)

....

c) -3(5c + 3)

....

d) -4(2d – 7)

....

e) -2(3e – 5)

....

These are more difficult. Multiply out the brackets, then collect the like terms. The first one is done for you:

Q3 a) 3(a – 1) + 2(a – 2)

3a – 3 + 2a – 4

5a – 7

b) 7(2b – 1) – (5b – 1)

....

c) -2(4c + 3) + 5(2c – 3)

....

d) -4(d – 5) – 3(2d – 3)

....

To solve these equations multiply out the brackets first. The first one is done for you.

Q4 a) 2(x + 5) = 6

2x + 10 = 6

2x = -4

x = -2

b) 3(x – 2) = 12

....

c) 4(2x + 3) = 20

....

6.7   *Questions on Substitution*

Don't forget the rules about negative numbers when you're substituting.

Q1   Find the value of $n + 5$ when n is:

a) 3 ....   **b)** 17 ....   **c)** 6.7 ....   **d)** -9 ....

    Work out the value of $3w + 2$ when w is:

e) 4 ....   **f)** 1.5 ....   **g)** -5 ....   **h)** $\frac{1}{4}$ ....

    If x = 4 and y = 2, find the value of:

i) $x^2 + y$ = ...   **j)** $3x - y$ = ...   **k)** $2x - 4y$ = ...

l) $\frac{6x}{2}$ = ...   **m)** $\frac{4x - 3y}{5}$ = ...   **n)** $x^2 \times y^2$ = ...

    Here is a formula $A = c - d^2$

o) Work out A when c = 50 and d = 6

          ....

    A formula for the cost in £ of repairing an Easiwash dishwasher is
    $C = 35 + 18 \times n$, where n = number of hours.

p) How much would it cost if it took 3 hours to repair? ....

q) How much would it cost if it took half an hour to repair? ....

r) If the repair cost was £80, how long did it take to repair it? ....

6.8   *Questions on Formulas*

Q1   Mrs. Jones works out the weekly pocket money for each of her children. She uses the formula:

Pocket money = Age in years × 20
(in pence)

Work out the pocket money for: **a)** Joe, aged 10 years ....

b) Paul, aged 8 years ....

c) Sara, aged 5 years ....

Q2   The formula to work out the cooking time for a turkey is:

Cooking time = Weight × 20 + 30
   (in mins)   (in pounds)

How long will it take to cook:

a) a 12 pound turkey ....
b) a 15 pound turkey ....

Q3   The formula to find the speed of a car is S = D ÷ T where S is the speed, D is the distance and T is the time. Use the formula to find the speed if:

a) D = 200 miles, T = 4 hours ....

b) D = 350 miles, T = 5 hours ....

Q4   The formula to work out the cost of hiring a carpet cleaner is C = 3d + 2 where C is the cost in £ and d is the number of days. Use the formula to find:

a) the cost of hiring the cleaner for 3 days ....

b) the cost of hiring the cleaner for a week ....

Q5   A formula to find the area of a rhombus is A = pq ÷ 2 where A is the area and p and q are the lengths of the diagonals. Use the formula to find

a) A when p = 6cm and q = 4cm. ....

b) A when p = 9.6cm and q = 6.4cm. ....

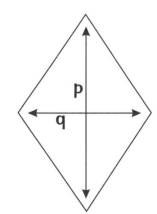

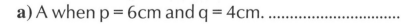

Show *every stage* of your working — write the *formula in words*, then again with the *numbers in*, then the *answer* — each bit could be worth something.

SECTION SIX — ALGEBRA

6.9   *Questions on Word Puzzles*

Q1   Write down the number I first thought of in each of these...

a) I multiply by 3 and then take away 5. The answer is 19.

My number was ....

b) I add 10 then double the number. The answer is 26.

My number was ....

c) I half it then add 9. The answer is 24.

My number was ....

d) I divide by 5 then add 12. The answer is 32.

My number was ....

e) I subtract 15 then divide by 3. The answer is 20.

My number was ....

Q2   Now try this...
I think of a number, multiply it by itself, add 5, divide by 2 then add 10. The answer is 25.

My original number was ....

Hint: the opposite to multiply by itself or squaring is $\sqrt{}$.

The trick is to work backwards from the answer, doing the opposite of each step (add instead of subtract, divide instead of times, that sort of thing).

6.10   Questions on One Step Equations

Q1   Solve these equations:

a) a + 6 = 20

b) b + 12 = 30

c) 48 + c = 77

d) 397 + d = 842

e) e + 9.8 = 14.1

f) 3+ f = 7

Q2   Solve these equations:

a) g – 7 = 4

b) h – 14 = 11

c) i – 38 = 46

d) j – 647 = 353

e) k – 6.4 = 2.9

f) l – 7 = -4

Q3   Solve these equations:

a) 4m = 28

b) 7n = 84

c) 15p = 645

d) 279q = 1395

e) 6.4r = 9.6

f) -5s = 35

Q4   Solve these equations:

a) $\frac{t}{3}$ = 5

b) u ÷ 6 = 9

c) $\frac{v}{11}$ = 8

d) $\frac{w}{197}$ = 7

e) x ÷ 1.8 = 7.2

f) $\frac{y}{-3}$ = 7

You've got to get the letter on its own (x = ...).
You can add, divide... well, anything really — but you
gotta do it to both sides or it'll all go horribly wrong.

6.11   *Questions on Two Step Equations*

Q1   Solve these equations:

    a) $3x + 2 = 14$ **b)** $5x - 4 = 31$

    c) $8 + 6x = 50$ **d)** $20 - 3x = -61$

Q2   Solve these equations:

    a) $\dfrac{x}{3} + 4 = 10$ **b)** $\dfrac{x}{5} - 9 = 6$

    c) $4 + \dfrac{x}{9} = 6$ **d)** $\dfrac{x}{17} - 11 = 31$

Q3   Solve these equations:

    a) $3\,(2x + 1) = 27$ **b)** $2\,(4x + 1) + x = 56$

    c) $5x + 3 = 2x + 15$ **d)** $2\,(x + 7) = 6x - 10$

These are just like the last page... only there's an extra step. But make sure you show both steps in your working — just in case you go wrong somewhere.

6.12   Questions on Trial and Improvement

Q1   Use the trial and improvement method to solve the equation $x^3 = 50$. Give your answer to one decimal place. Two trials have been done for you.

Try $x = 3$ $x^3 = 27$ (too small)
Try $x = 4$ $x^4 = 64$ (too big)

....

Q2   Use the trial and improvement method to solve these equations. Give your answers to one decimal place.

a) $x^2 + x = 80$

....

b) $x^3 - x = 100$

....

**Show all the numbers you've tried, not just your final answer...
or you'll be chucking away easy marks.**

6.13 Questions on Real Life Equations

Write an equation which describes each of the situations given. Solve each equation.

Q1 I have 3 bags of sweets, each with the same number of sweets (call this s). I eat 7 sweets. I now have 29 sweets left. How many sweets were in each bag (s) to start with?

....

Q2 The number of peanuts in a bag (call this p) is shared between 5 friends, each friend gets 6 peanuts with 3 peanuts left over. How many peanuts were in the bag (p)?

....

Q3 A rectangle is x cm long, the height of the rectangle is 5cm more than this. The perimeter of the rectangle is 40cm. Find the length (x).

....

Q4 Tom's age (call this t) when multiplied by 6 is 10 less than his father's age. His father is 28. How old is Tom (t).

....

Don't forget to check your answers by putting the numbers back in. If it doesn't work you've messed up somewhere, so try again.

Exam Questions Without a Calculator

Q1 Find the cost of:

a) 10 shirts each costing £16.50 ....

b) 100 books each costing £2.99 ....

Q2 At midnight the temperature was 1°C.

a) By 6 a.m. it has fallen by 5°. What is the temperature now? ....

b) By 10 a.m. the temperature has risen again to 0°C.

How many degrees has it risen since 6 a.m.? ....

Q3 Fifteen members of the lottery syndicate pay £26 each year.

a) How much do they pay altogether? (Show your working out)

b) They win the jackpot and each member gets £8523. How much did they win altogether?

Q4 A piece of ribbon is 354cm long. It is cut into lengths 16cm long.

a) How many lengths can it be cut into?

....

b) How much ribbon will be left over? ....

Q5 Show how to estimate the answer to 582 × 34.

....

....

Q6 How many 52 seater coaches are needed to take 384 fans to a football match?

Exam Questions Without a Calculator

Q7

1	2	3	4	5	6	7	8	9	10	11	12
13	14	15	16	17	18	19	20	21	22	23	24
25	26	27	28	29	30	31	32	33	34	35	36

From the numbers in the box write down:

a) all the multiples of 6 ....

b) all the factors of 24 ....

c) the square of 5 ....

d) the cube of 3 ....

e) the prime numbers between 21 and 30 ....

f) the square root of 25 ....

g) a multiple of both 5 and 7 ....

Q8 Here is a rule for working out a number sequence.

Starting number ⟹ Multiply by 2 ⟹ Add 5 ⟹ End number

a) Use the rule to continue the sequence for four more terms:

Starting Number	1	2	3	4	5	6
End Number	7	9				

b) What sort of number are all your answers? ....

c) What sort of numbers are these? 1, 4, 9, 16, 25, 36 ....

d) And these? 1, 3, 6, 10, 15, 21 ....

Q9 a) What is the perimeter of this rectangle? ....

b) What is its area? ....

7cm

12cm

Q10 There are 150 pupils in Year 10. 2/5 of the pupils are boys.

a) Work out how many boys there are? ....

b) How many girls are there? ....

Exam Questions Without a Calculator

Q11 John scored 50% on a Maths test. The total number of marks were 150. How many marks did John score? ....

Q12 Measure this line.

Write your answer in: **a)** centimetres .... **b)** millimetres ....

Q13 Look at this number pattern.

Use the pattern to work out the next two lines.

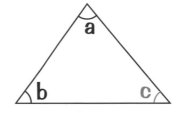

$$7 \times 7 \quad = 49$$
$$67 \times 67 \quad = 4489$$
$$667 \times 667 = 444889$$

....

....

Q14 Angle a is 65° and angle b is 62°.

What is the size of angle c?

....

....

(The triangle is not drawn to scale so you can't measure the angle.)

Q15 Books cost £3.95 each.

a) How many can you buy with £20? ....

b) How much change would you get? ....

Q16 Fill in the spaces:

	Fraction	Decimal	Percentage
a)	½	0.5	....
b)	....	0.7	....
c)	¾	....	75%
d)	3/100	....	....

Q17 Jane works for 6 hours 15mins. She is paid £4 an hour.

a) What is her total pay? ....

b) How many minutes does she work?

....

Exam Questions Without a Calculator

Q18

Think of a number, double it and add 3

My answer was 21   Paul

The number I thought of was 12   Philip

a) What was the number Paul thought of? ....

b) What answer did Philip get? ....

Q19

Eiffel Tower	Big Ben	Statue of Liberty	Acropolis
984 feet	320 feet	306 feet	500 feet

a) Write the height of the Eiffel Tower to the nearest 10 feet. ....

b) Write the height of the Eiffel Tower to the nearest 100 feet ....

c) How much higher is the Acropolis than the Statue of Liberty? ....

d) When their heights are written to the nearest 100 feet, which two buildings are the same height?

....

e) Which building is about 100 metres high? (3 feet is approximately 1 metre)

....

f) Put the heights in size order, smallest first ....

....

Exam Questions With a Calculator

You may use a calculator for these questions.

Q1 Ruth gets on a train that leaves Manchester at 1.15p.m. and reaches Lancaster at 2.50p.m.

a) How long does the journey take? ....

Ruth gets off the train at Lancaster and goes to catch a train to Morecambe. The train leaves at 3.05p.m.

b) How long has she to wait? ....

c) The journey to Morecambe takes 12 minutes. If the train leaves on time what time will it reach Morecambe?

....

d) How long has Ruth's journey taken, including waiting time?

....

Q2 To form a sequence of numbers the rule is:

"Multiply the previous number by 2 then add 1"

a) Write down the next three numbers of this sequence

3   7   15   ....

b) List the numbers of the sequence which are multiples of 3:

....

c) List the first 3 numbers in the sequence which are Prime numbers:

....

Q3 Solve these equations:

a) $3m + 7 = 19$ **b)** $7y + 3 = 21 + y$ **c)** $4(u + 2) = 32$

....

Exam Questions With a Calculator

Q4   Below is the list of counting numbers from 1 to 30.

1	2	3	4	5	6	7	8	9	10
11	12	13	14	15	16	17	18	19	20
21	22	23	24	25	26	27	28	29	30

a) Make a list of the prime numbers:

....

....

b) Make a list of the square numbers:

....

....

c) Make a list of the cube numbers:

....

....

d) Make a list of the triangular numbers:

....

....

e) List the set of four prime numbers which add up to 50:

....

....

Exam Questions With a Calculator

Q5 a) Look at the following shapes and write down the names of them:

 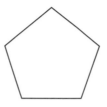 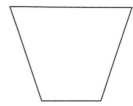

Name ....   Name ....   Name ....

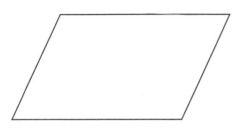 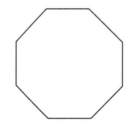 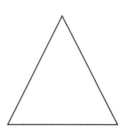

Name ....   Name ....   Name ....

b) Two of the shapes have only one line of symmetry. Draw the line of symmetry on each shape.

c)

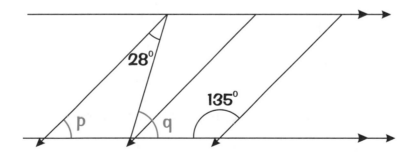

Calculate the value of angle 'p' and give a reason for your answer:

p = ...° because ....

Calculate the value of angle 'q' and give a reason for your answer:

q = ...° because ....

Exam Questions With a Calculator

Q6 Picnic tables.

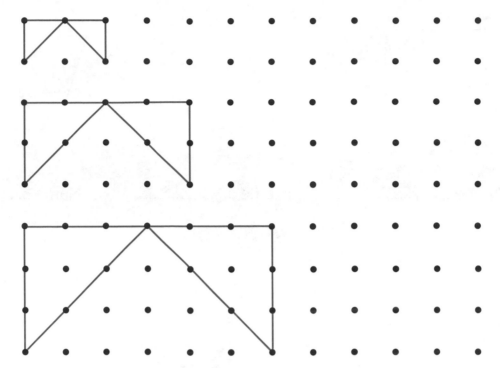

a) Fill in the table below for the first five picnic tables.

Picnic table number	Number of dots
1	5
2	
3	
4	
5	

b) Write in words a rule which links the table number to the number of dots.

....

c) Write the rule you have found as a formula. Use 't' for the table number and 'd' for the number of dots.

Answer ....

d) Use your formula to find how many dots there are in the 20 th picnic table.

Answer ....

Exam Questions With a Calculator

Q7   A probability line is shown below. The arrow shows the probability of selecting a red card from a pack of 52cards

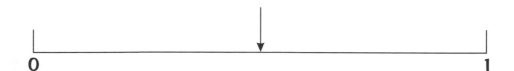

0   1

a) Put an arrow on the line to show the probability of selecting an ace.
Label it 'a'

b) Put an arrow on the line to show the probability of not selecting a Heart.
Label it 'b'

Q8   A school runs a raffle to raise money and collects £420 altogether. The first three prizes are a percentage of the money collected as follows:

1st prize – 20% ,   2nd prize – 15% ,   3rd prize – 10%

a) How much money is given to each of the first three prize winners?

1st prize = £..........

2nd prize = £..........

3rd prize = £..........

b) $\frac{3}{8}$ of the total money collected is spent on new library books. What amount is spent on library books?

Answer £ ....

c) How much money is left over?

Answer £ ....

Exam Questions With a Calculator

Q9 23822 people attended a pop concert. Round this number

   a) To the nearest hundred. Answer.............

   b) To the nearest thousand. Answer.............

Q10 'Purplex' paint is made by mixing red and blue paint in the ratio 5 : 3.

   a) How much of each colour would you need to make:

     i) 24 litres of Purplex? Answer ....

     ii) 4 litres of Purplex? Answer ....

   b) If you used 15 litres of blue, how much red paint would you need to mix with it to make Purplex?

     Answer ....

Q11 Here is a triangle:

8.4cm   8.4cm   7.5cm   64°   7.4cm

   a) Work out the perimeter of the triangle:

     Answer ....

   b) Work out the area of the triangle:

     Answer ....

   c) Work out the other two angles of the triangle:

     Answer .... and ....

Q12 a) The probability that I will catch my train is **5/8**.

   What is the probability that I will miss it? ....

   b) The probability that Useless Wanderers will win their next match is 0.1 and the probability they will draw is 0.2. What is the probability that they will lose?

     ....

Q13 A pupil is chosen at random from a class of 14 boys and 16 girls. What is the probability that a boy is chosen?

....

Exam Questions With a Calculator

Q14 The graph shows the journeys of Jim and Fiona. They both travel by car on the same road but in opposite directions.

a) How far is Rutterton from Woodster? ....

b) At what time did they both set off? ....

c) At approximately what time did they pass each other? ....

d) Estimate how far were they from Woodster when they passed each other? ....

e) Jim stopped for a break on the way to Woodster. How long did he stop for? ....

f) What was Fiona's average speed for the journey? ....

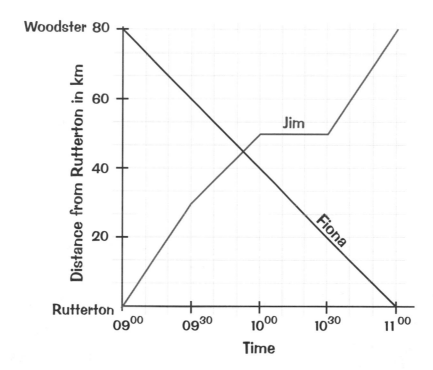

Q15

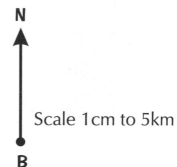

Scale 1cm to 5km

a) What is the actual distance from A to B? ....

b) What is the bearing of B from A? ....

c) What is the bearing of A from B? ....

Exam Questions With a Calculator

Q16 Here are Julie's and Karen's times in swimming trials for the 25m race over the last five weeks: (all times in seconds).

Julie	17.1,	15.8,	15.3,	14.8,	15.0
Karen	16.2,	15.7,	15.4,	15.1,	15.1

a) Julie's mean average time is 15.6 seconds. What is Karen's mean average time?

....

b) Work out the range of times for each girl.

Julie ....

Karen ....

c) Who would you choose to represent the school at the district gala? Give a reason for your choice.

....

....

....

Q17 The pie chart shows how 600 pupils usually travel to school.

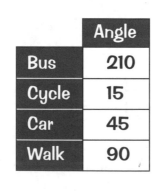

	Angle
Bus	210
Cycle	15
Car	45
Walk	90

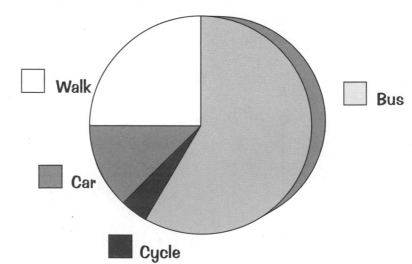

Calculate how many pupils are in each sector of the pie chart.

walk............ bus............. car.............. cycle............

MFW41